Aspects of

EAST ANGLIAN STE

VOLUME TWO

ON EASTERN BRANCH LINES AND

INTRODUCTION

Not so long ago it was possible to travel almost anywhere in East Anglia by rail, however remote the destination might have been; after all, railways were built to serve the community. Problems arose, where branch lines were concerned, when dwindling patronage and a dramatic increase in road transport sucked the lifeblood from rural byways, which then fell prey to the "Beeching" cuts.

Many of these lines not only served but became part of a community; local interest rekindled when the closure notices went up was for the most part too late and ineffective, leaving a majority of East Anglia's provincial areas trainless.

The ones that survived have been "tailored" for today's needs, trackwork and property kept to minimum; a few, with electrification, have once again become popular.

This album does not set out to document every branch – far from it – the idea is to present a selection of lines, from the busy suburban routes around London to the delightfully timeless Mid Suffolk Light Railway (a pre Beeching casualty), where it was possible to travel in ex-G.E.R. six-wheel coaches well after nationalisation. Lingering where photographs permit at locations of special interest.

Also included are scenes on the M.&G.N., a cross-country line of great character once well represented in East Anglia, in its hey-day a very efficient organisation even boasting a railway works at Melton Constable. The whole system was another early victim of misfortune; only a small section survives, operated jointly by B.R. and the North Norfolk Railway, who operate perhaps the most scenic section between Sheringham and Holt. Earlier attempts to preserve larger parts of "the joint" were thwarted on financial grounds. A bold "eleventh hour" bid way back in 1965 only just ahead of "cutters' torches" secured what has become one of East Anglia's most endearing preserved lines. *John D. Mann, Frinton-on-Sea, 1991*

ACKNOWLEDGEMENTS

Compiling this album became a labour of love and is the result of several years' research for suitable photographs. I am indebted to everyone who came forward with offers of material, many of whom loaned negatives.

I should like to thank author and photographer Peter Hay for the opportunity of viewing his entire East Anglian collection, and for the subsequent very enjoyable hours spent in the darkroom. I extend thanks to Tony Bennett, Jack Kirke, Bruce Nathan and Philip Kelley for similar services rendered. Les Peters also spent considerable periods of time in his darkroom on my account producing some quite outstanding scenes from the early 1950s. I must also mention Michael Mensing, Ernest Alger, Alan Pike, Peter Snell (F. Church collection) and the Lavenham Press Ltd. for their assistance. Finally, I was particularly pleased to secure the help of John Dewing, who kindly called personally to place prints at my disposal.

FRONT COVER
HAUGHLEY JUNCTION – July 5th 1952. Modern electric trains now speed through this former Suffolk Junction. During the summer of 1952 the scene was very different. Reflecting a far more leisurely age of East Anglian rail travel, Class J15 No 65447 is pictured after arrival from Laxfield (Mid Suffolk Light Railway). The line closed completely on July 26th 1952. (*Photo* – L. R. Peters)

INSIDE BACK COVER (upper)
ASHDON HALT – August 12th 1961. Steam had largely disappeared from the Audley End–Bartlow branch when this view was taken. Ashdon Halt however retained its distinctive wayside atmosphere so typical of East Anglian branch lines until closure in 1964. An old G.E.R. coach body of 1883 sufficed as a waiting room and throughout its existence was unstaffed. The halt was situated near the top of a 1 in 80 gradient, almost unbelievable for Essex. (*Photo* – J. A. C. Kirke)

BACK COVER
NEAR SOUTH LYNN – August 1st 1938. The M.&G.N. in L.N.E.R. days. A King's Lynn–Peterborough (via Sutton Bridge) train scampers along behind Class D52 No 011; the loco had only four more years in service. (*Photo* – J. G. Dewing)

INSIDE FRONT COVER
ALDEBURGH – May 5th 1956. Class J15 No 65467 draws to a halt under the all-over roof at the attractive terminus at Aldeburgh on a sunny spring day. The branch closed completely east of Leiston during September 1966 and the station is now demolished. Freight traffic for Sizewell continues to Leiston. (*Photo* – P. J. Kelley)

INSIDE BACK COVER (lower)
HOPTON-ON-SEA – July 27th 1957. A scene on the approach road to the station on the arrival at 5.56pm of the 3.00pm (S.O.) Holiday Camps Express from Liverpool Street. Children are waiting to barrow holiday makers' luggage to the camps. This train ran non-stop to Lowestoft (Central) in 2½ hours, from where, after reversal, it was booked through to Yarmouth (South Town) calling at Corton, Hopton and Gorleston (Norfolk and Suffolk Joint line). (*Photo* – E. Alger)

THIS PAGE
CHAPPEL & WAKES COLNE – May 21st 1956. The 32 arch viaduct spanning the Colne Valley was by far the most impressive structure on the former G.E.R. Seven million bricks were used in its construction, and it is still in daily use carrying trains on the truncated Stour Valley line from Marks Tey to Sudbury. On this occasion Class J15 No 65478 crosses the viaduct with a Colne Valley line train. The loco is fitted with a side window and back cab, specifically designed for crew protection over this route. (*Photo* – P. Hay)

Published by South Anglia Productions, 26 Rainham Way, Frinton-on-Sea, Essex CO13 9NS. Tel: 0255 677965.
ISBN 1 871277 11 6
Printed in England by The Lavenham Press Ltd.

STRATFORD MARKET

The fireman of J69/1 No 68579 has obviously just "put a few shovelfuls round" while running bunker first on a North Woolwich–Palace Gates train on July 25th 1956. J69s were often called upon to perform some pretty heavy passenger work at times when suitable locomotive power was unavailable. No 68679 lasted until January 1960. (*Photo* – P. Hay)

BUCKHURST HILL

Class J15 No 65450 (pilot) and No 65476 disturb the tranquil suburbia of Buckhurst Hill with the 9.08am excursion from Loughton to Eastbourne on July 19th 1953. This was the first steam-hauled train on the line since the Central Line extension to Epping and presents a quite magnificent spectacle. (*Photo* – P. J. Kelley)

ENFIELD TOWN

Class F4 No 7233 stands with a rake of articulated coaching stock on September 24th 1938. These G.E. veterans famous for their appetite for coal were widely distributed throughout the Eastern Counties. Renumbered No 67166 in 1949, the loco lasted only until 1951. Note the injector overflow under the cab steps. *(Photo* – I. G. Dewing)

LOUGHTON

Prior to the Central Line extension on former G.E.R. tracks, Class F5 No 7210 (B.R. No 67210 from 1950) leaves Loughton on a glorious spring day, April 20th 1949. (*Photo* – J. G. Dewing)

WICKHAM BISHOPS

Above: Yet another East Anglian branch line no longer in existence. A Witham–Maldon East train, hauled by Class F5 No 67189 crosses the trestle bridge at Wickham Bishops on May 21st 1956. Although the line closed during 1966, the bridge remains largely intact today. (*Photo* – P. Hay)

BRIGHTLINGSEA

Below: A chilly easterly wind could have been blowing at Brightlingsea on March 14th 1952 as station staff prepare to dispatch a branch train to Wivenhoe. The tail lamp is already being walked along the platform for repositioning on the rear coach. The locomotive, Class J15 No 65432, has run round and recoupled; note the ample coal supply. The branch closed completely in June 1964. (*Photo* – B. I. Nathan)

FRINTON-ON-SEA

A scene which has changed little over the years, the famous "Frinton Gates" are still in use today. Overhead catenary is already in position for the start of electric services (March 1959). The Clacton–Walton–Colchester line was designated a "test bed" for the now standard 25Kv system; its success brought an early end to steam, although for a short period it was not uncommon to witness brand new electric units working alongside elderly J15s. Running tender first, No 65468 enters Frinton-on-Sea with the Walton-on-Naze–Thorpe-Le-Soken portion of a Liverpool Street train. The passing loop (left) was removed during the 1960s. The date is July 29th 1958. (*Photo* – F. Church, courtesy E.N.E.G.)

FELIXSTOWE TOWN

Above: The purposeful Thompson L1 tanks were once a familiar sight on the Felixstowe branch. In this photograph No 67703 is leaving (town) station for Felixstowe (Beach) after reversal, February 22nd 1958. (*Photo* – B. I. Nathan)

MARKS TEY

Below: The five railbuses supplied by Wagen- und Maschinenbau of Germany in 1958 for East Anglian branch work were notoriously unreliable. On this occasion No E79964 is being towed by Class B17/6 No 61651 "DERBY COUNTY" through the Stour Valley platform at Marks Tey, probably en route to Cambridge for attention. The magnificent bracket signal is still intact, and notice also that the scene is completely bereft of overhead wiring. It is March 21st 1959. (*Photo* – M. Mensing)

MARKS TEY

The Stour Valley line was a typical G.E. branch; it meandered through rolling cornfields and picturesque villages, but was an important cross-country route, especially during wartime. In this superb scene, Class J15 No 65457 is awaiting departure from Marks Tey at the head of a Cambridge bound train. With 160lbs "on the Clock", safety valves are lifting. Plenty of steam will be needed for the undulating 3½ miles to Chappel & Wakes Colne. The date is May 28th 1958. (*Photo* – J. A. C. Kirke)

GREAT TEY

Class D16/3 No 62610 makes steady progress along the Stour Valley line shortly after leaving Marks Tey with a Cambridge–Colchester train, May 21st 1956. (*Photo* – P. Hay)

CHAPPEL & WAKES COLNE

Once an important through route, the Stour Valley line has led a chequered existence over the last few decades. A large proportion of the line closed in 1967 and closure threatened the remaining section periodically. This seemingly timeless view was recorded in less troubled times on May 21st 1956. A Cambridge–Colchester train, headed by Class J15 No 65390, leaves Chappel & Wakes Colne, crossing the viaduct (receiving attention, note scaffolding left of picture). The station just visible in the distance is now the East Anglian Railway Museum site. (*Photo* – P. Hay)

CHAPPEL & WAKES COLNE

A closer look at Chappel & Wakes Colne station, May 21st 1956. Thanks to the efforts of the East Anglian Railway Museum this classic Essex country junction remains much the same today. In this fascinating view, Class J15 No 65438 (ex-Colne Valley line) is arriving "wrong platform" so that the train from Cambridge via Sudbury can be accepted, while the Colne Valley train is still in the station. Note the tablet exchange. (*Photo* – P. Hay)

SUDBURY

Above: B.R's last 2–4–0, Class E4 No 62785, is leaving Sudbury with a Cambridge–Colchester train and making one of its last runs before withdrawal for preservation on a misty autumn day, October 1958. After many years of dereliction the station buildings were finally demolished during 1990. The footbridge, resited, has found new life at Chappel & Wakes Colne. (*Photo* – J. G. Dewing)

HALSTEAD

Below: Although situated very firmly in G.E. territory the Colne Valley line remained independent until absorbed by the L.N.E.R. in 1923; it was also sadly underphotographed, especially during its latter stages. On this occasion passengers have spilled out of an R.C.T.S. special calling at Halstead during an extensive tour on August 10th 1958. The loco for this section of the journey is Class J15 No 65440, having taken over at Marks Tey. (*Photo* – A. E. Bennett)

STOKE BY CLARE

North of Sudbury, the Stour Valley line closed completely in March 1967. Reviving memories of the lovely section between Long Melford and Haverhill, Class J15 No 65472 leaves Stoke by Clare on a sunny summer's day, August 30th 1958. The loco had only sixteen months' service left, being withdrawn in December 1959. Somehow a J15 always looked just right on an East Anglian branch. (*Photo* – J. G. Dewing)

HAVERHILL

It was always exciting being a passenger in the first coach behind the locomotive, especially a J15 running tender first, exhaust booming through the tall chimney. On this train, however, a Colne Valley service leaving Haverhill, a first class ticket would be required. The date is July 5th 1952. (*Photo* – L. R. Peters)

BARTLOW

Bartlow Junction was a rather curious station; situated in an isolated setting it boasted two signal boxes before the L.N.E.R. regime took over in 1923. Primarily its function was to accommodate trains from Audley End and Saffron Walden (by means of a simple wooden platform) and Stour Valley services. With G.E. signalling very much in evidence, Class B12/3 No 61557 restarts a train bound for Colchester during the summer of 1955.

(*Photo* – L. R. Peters)

BARTLOW

Another B12/3, No 61580 sets out for Cambridge on August 30th 1958, with the train crew looking decidedly cheerful. Note the tablet about to be exchanged and the platform oil lamps. (*Photo* – J. G Dewing)

SAFFRON WALDEN

May 1935, and Class G4 No 8139 in L.N.E.R. Black livery stands at Saffron Walden with a train for Audley End. The scene has a distinct G.E. flavour highlighted by the six-wheel brake/comp next to the engine and the ground signal in the foreground.

(*Photo* – J. G. Dewing)

SAFFRON WALDEN

A superb photograph which conveys immediately the flavour of the Saffron Walden branch during the early "fifties". Class C12 No 67375, at this time allocated to Cambridge, takes water on June 3rd 1951. Once a regular sight on the line during the "forties", C12s were displaced by G5 tanks after the introduction of Push and Pull trains in July 1951. (*Photo* – L. R. Peters)

SAFFRON WALDEN

Rooks' nests stand out clearly on the skyline above the station as Class J20 No 64696 is pictured on an engineers' train on April 22nd 1956. These strong ex-G.E. 0–6–0s were often employed on heavy goods work and trains such as this one. Four survived at March until 1962. (*Photo* – L. R. Peters)

BOTTISHAM

Class E4 No 62785 pulls out of Bottisham with a Mildenhall–Cambridge train, May 5th 1958. There is plenty for the railway modeller here, the lower quadrant semaphore being of special interest. Note also the early crops growing in an allotment near the platform; most railwaymen were keen growers of home produce, an obvious throwback to wartime, when it was of special importance to be self-sufficient. Photo – J. G. Dewing)

BURY ST EDMUNDS

Complete with headboard, Class J15 No 65391 stands on Bury St. Edmunds shed prior to working over the Bury–Thetford branch on the last day of passenger services, June 6th 1953. Freight continued until 1960. 65391 is an 1890 vintage J15 and was over sixty years old when this photograph was taken; she lasted until 1958. (*Photo* – L. R. Peters)

BARNHAM

On the last day of passenger services, Class F6 No 67237 calls at Barnham on the Bury St. Edmunds–Thetford branch, June 6th 1953. The train is augmented to four coaches to accommodate extra passengers paying their last respects. The close proximity of Barnham camp and bomb dump often necessitated the fitting of spark arresters to loco chimneys.

(*Photo* – L. R. Peters)

ALDEBURGH

Prompt to the minute the branch train from the Ipswich–Lowestoft line at Saxmundham arrives at the gas lit Aldeburgh terminus. The layout here was traditional, with one platform, a run-round loop and a small goods yard beyond. The signal box and water tower complete the picture. The loco is Class F6 No 67230, carrying a 32B (Ipswich, Aldeburgh) shedplate. During the summer of 1949/50 an L.M.S. 2–6–2T was loaned from Bangor by the chairman of the Railway executive, after a weekend visit to the branch when the unfortunate rostered engine failed. (*Photo* – P. Hay)

MENDLESHAM

A view on the Mid Suffolk Light Railway shortly before closure, July 5th 1952. The fireman of Class J15 No 65447 has opened the crossing gates near Mendlesham. The "Middy", as it was affectionately known, served mainly remote villages along its route, and became an early candidate for closure. The end came on July 26th 1952. (*Photo* – L. R. Peters)

WYMONDHAM

A nicely turned out Class D16/3 No 62620 gets under way having negotiated the junction at Wymondham, August 25th 1953. The train, which is very "mixed", is bound for Dereham. Note the M.&G.N. Whittaker tablet catcher fitted to the tender, and the 32G Melton Constable shedplate. (*Photo* – L. R. Peters)

HUNSTANTON

The terminus at Hunstanton (pronounced Hunston) looks deserted as an unidentified Class D16/3 prepares to leave on a King's Lynn Stopping train during June 1960. It is hard to comprehend that this view has completely disappeared since closure in 1969. The line was immortalised by the late John Betjeman in a B.B.C. East programme which is still screened periodically. (*Photo* – A. J. Pike)

WISBECH

A pleasant reminder of the Upwell goods leaving Wisbech on August 9th 1937. The loco is Class J70 No 7136 and is fitted with side skirts covering the motion and a cowcatcher, a requirement for working over public roads. 7136 became B.R. No 68217 and was withdrawn in 1953.

(*Photo* – J. G. Dewing)

MELTON CONSTABLE

Having worked through from Cromer, Class B17/6 No 61654 "SUNDERLAND" is on the turntable road. Melton Constable was known as the "Crewe of North Norfolk", being the centre of operations for the M.&G.N. It is a bright February 28th 1959, and major closure of the M.&G.N. system is imminent. Close inspection of this wonderful view will reveal a multitude of fine details, including a mother showing a piece of railway history to her young offspring.
(*Photo* – A. E. Bennett)

MELTON CONSTABLE

The last 9.02am ex-Yarmouth through train leaves Melton Constable, February 28th 1959. Note the ex-G.E.R. restaurant car next to the locomotive. Enthusiasts are well represented, and the signalman shades his eyes to witness an historic moment. (*Photo* – A. E. Bennett)

SOUTH LYNN

The M.&G.N. was a regular haunt for Ivatt Class 4 2–6–0s; despite their rather ungainly appearance they were well suited to the line's requirements. No 43155 puts up a fine display of smoke and steam at South Lynn on August 1st 1958. (*Photo* – L. R. Peters)

SOUTH LYNN

Class 4F No 44401 eases forward from a holiday excursion at South Lynn, on August 16th 1952, and prepares to hand over to Ivatt Class 4 No 43158. The North Norfolk resorts became popular thanks to the M.&G.N.; South Lynn in particular became very busy during summer weekends. (*Photo* – L. R. Peters)

SOUTH LYNN

Ivatt Class 4 No 43158 "has the road" and departs from South Lynn with excursion M64, August 16th 1952. The Whittaker tablet catcher can be observed in the tender recess. Although the M.&G.N. was noted for its long stretches of single line, smart working could be maintained with this equipment, exchange being possible at speeds up to 50mph.

(*Photo* – L. R. Peters)

SOUTH LYNN

Freight traffic from the fertile fens west of King's Lynn was a major revenue earner for the M.&G.N., in fact almost 90% carried being of agricultural origin. South Lynn was opened for goods traffic in 1863 over 20 years before the passenger station. In this view Class J11 No 64420 fusses through on August 1st 1953. (*Photo* – L. R. Peters)

SOUTH LYNN

A study of the South Lynn–King's Lynn Push and Pull train, August 1st 1953. The locomotive is Class C12 No 67374. (*Photo* – L. R. Peters)

SOUTH LYNN

Three 4Fs Nos 44423–43937–44412 await turns of duty at South Lynn during a busy summer weekend; the date is August 1st 1953. Photo – L. R. Peters)

SOUTH LYNN

Another view of the South Lynn–King's Lynn Push and Pull train arriving at South Lynn behind Class C12 No 67374 on August 1st 1953. These handsome passenger tank engines were first introduced by H. A. Ivatt in 1898 – this example was built in 1899 and lasted until 1958 (push and pull fitted 1948). (*Photo* – L. R. Peters)

SOUTH LYNN

A photograph with a certain air of fir.ality. Just two weeks before closure of the M.&G.N. system the 1.40pm King's Lynn–Peterborough train leaves South Lynn on February 14th 1959. (*Photo* – A. E. Bennett)

SHERINGHAM

British Rail services into Sheringham were transferred to a new halt erected to the east of station road level crossing in January 1967. This view shows a Melton Constable bound D.M.U. arriving past an Ivatt Class 4MT on February 28th 1959 (the M.&G.N.'s last day). The station now forms the starting point for journeys on the North Norfolk Railway.

(*Photo* – A. E. Bennett)

RUNTON

Class D16/3 No 62581 arrives at Runton West Junction from the Sheringham (M.&G.N.) direction with a passenger train for Cromer (Beach) on June 16th 1952; the signalman is ready to receive the tablet. The line (left) briefly joined the Norfolk and Suffolk Joint Railway, before curving onto the G.E.R. at Roughton Road Junction, en route to North Walsham.

(*Photo* – L. R. Peters)

CROMER

A lovely photograph of Cromer Beach shed, illustrating its extremely rural setting. Basking in the summer sunshine of June 15th 1952 are Class F6 Nos 67224 and 67225. The imposing church tower of Cromer can just be seen through the heat haze. Situated behind the loco (shed road) is a truck containing a coal supply. As allocations at these sub-sheds were coaled by hand, smaller engines were obviously more popular. (*Photo* – L. R. Peters)

AYLSHAM

A brand new railway now occupies this site, the 15in gauge Bure Valley line, constructed on the trackbed of the former standard gauge branch from Wroxham. The scene is Aylsham on a damp October 8th 1960, and Class B12/3 No 61572 is calling with a special train organised by the M.&G.N. Preservation Society. Such was the competition between the G.E.R. and the M.&G.N. in Norfolk that many towns had two stations; the M.&G.N. used Aylsham North. The train is particularly interesting. Not only is it being hauled by the very last B12 to remain in service, but it was also claimed to be the first passenger train over the newly laid Thelmelthorpe spur, and later was the last train to traverse the Waveney Valley line.

(*Photo* – A. E. Bennett)

GUNTON

An unidentified Class N7 tank coasts along with the 7.57am train from Lowestoft (Central) to Yarmouth (South Town) and crosses the embankment near Gunton on July 27th 1957. The right hand side of photo now forms part of the Pleasurewood Hills theme park.

(*Photo* – E. Alger)

GORLESTON-ON-SEA

Class D16/3 No 62613 starts the 09.00 (S.O.) Gorleston-on-Sea to Norwich and York. This train would traverse the Norfolk and Suffolk Joint line to Lowestoft (Central), where reversal would be carried out. Gorleston Holiday Camp is to the left of the photograph and was connected by a path directly to the station. No dcubt most of the passengers would be returning home after holidays at the camp. (*Photo* – E. Alger)

HOPTON-ON-SEA

The "Holiday Camps Express" (3.00pm S.O.) Liverpool Street to Yarmouth (South Town) arrives at Hopton-on-Sea on July 27th 1957. The locomotive in charge is Class D16/3 No 62570. (*Photo* – E. Alger)

CORTON

The 2.00pm Yarmouth (South Town)–Lowestoft (Central) approaches Corton hauled by Class D16/3 No 62613 with steam to spare, April 6th 1957. The Norfolk and Suffolk Joint Railway was operated by the L.N.E.R. from 1936, and was incorporated into British Railways from January 1948. The line closed in 1967 (goods), a passenger service surviving until 1970.

(*Photo* – E. Alger)

CORTON

Another view on the Norfolk and Suffolk Joint. Class N7/3 No 69679 coasts past Corton church with the 2.01pm from Yarmouth (South Town) to Lowestoft (Central) on July 27th 1957. The loco is fitted with M.&G.N. Whittaker tablet exchange apparatus (recess in side tank) and is also push and pull fitted.
(*Photo* – E. Alger)

Above: Remarkably four Class J15s remained in B.R. stock in September 1962. No 65462 and the last B12/3, No 61572, were purchased by the M.&G.N. Joint Railway Society. Both engines resided in the old steam depot at March in store until 1967, when finally they left B.R. metals for Sheringham and eventual restoration. It was to be 1977 before 65462 entered service on the North Norfolk Railway. In this view we see the loco simmering gently between duties at Beccles on April 6th 1956; at this time she carried the medium length chimney and a tender cab. Being a 1912 vintage J15 No 65462 was one of the younger members of a once numerous G.E. class. (*Photo* – P. Hay)

THE SURVIVORS

Below: In complete contrast, the only Class N7 to be saved, No 69621, was purchased privately and spent many years in store at Neville Hill Depot, Leeds. In 1973 it was towed down to Colchester for restoration at the East Anglian Railway Museum at Chappel & Wakes Colne. Sixteen long years passed before the familiar sound of a Westinghouse pump was heard again, the engine returning to active life in the summer of 1989. On April 28th 1962, 69621 was employed on an L.C.G.B. special "The Great Eastern Suburban Railtour", and is pictured taking water at Palace Gates.

(*Photo* – J. A. C. Kirke)

EYFS Early Adopters Development Matters

Non Statutory EYFS Guidance for Early Adopter Schools

To purchase a copy please visit:
www.TheNationalCurriculum.com
or scan this code to take you there:

Published by: Shurville Publishing

Also available for download at https://www.gov.uk/government/publications

ISBN: 9781838150211

Development Matters

Non-statutory curriculum guidance for the early years foundation stage

September 2020

Contents

Introduction

No job is more important than working with children in the early years.

Development Matters has been written for all early years practitioners, for childminders and staff in nurseries, nursery schools, and nursery and reception classes in school. It offers a top-level view of how children develop and learn. It guides, but does not replace, professional judgement.

The guidance can also help you to meet the requirements of the *statutory framework for the early years foundation stage*. The framework sets out the three prime areas of learning that underpin everything in the early years:

- communication and language
- physical development
- personal, social and emotional development

The four specific areas help children to strengthen and apply the prime areas:

- literacy
- mathematics
- understanding the world
- expressive arts and design

All of those areas of learning are connected together. The characteristics of effective teaching and learning weave through them all. That's because children in the early years are becoming more powerful learners and thinkers. These characteristics develop as they learn to do new things, acquire new skills, develop socially and emotionally, and become better communicators.

Development Matters sets out the pathways of children's development in broad ages and stages. However, the actual learning of young children is not so neat and orderly. The main purpose of these pathways is therefore to help you assess each child's level of development. Accurate assessment helps practitioners to make informed decisions about what a child needs to learn and be able to do next.

The document is not a tick list for generating lots of data. You can use your professional knowledge to help children make progress without needing to record lots of next steps. Settings can help children to make progress without generating unnecessary paperwork.

Examples of effective practice mentioned early on are often relevant for older children. For example, the communication and language section says that 'babies and toddlers thrive when you show a genuine interest in them, join in and respond warmly.'

Of course, this is also true for children of all ages.

The guidance can help you check that children are secure in all the earlier steps of learning before you look at their 'age band'. Depth in learning matters much more than moving from one band to the next or trying to cover everything. A child's learning is secure if they show it consistently and in a range of different contexts. For example, it is important to give a child many opportunities to deepen their understanding of numbers up to five. There is no value in rushing to 10.

The observation checkpoints can help you to notice whether a child is at risk of falling behind in their development. You can make all the difference by taking action quickly, using your professional judgement and your understanding of child development. By monitoring the child's progress more closely, you can make the right decisions about what sort of extra help is needed. Through sensitive dialogue with parents ('parent' is used throughout this document to refer to parents, carers and guardians), you can begin to understand the child better and also offer helpful suggestions to support the home learning environment.

When children are at earlier stages of development than expected, it is important to notice what they enjoy doing and also find out where their difficulties may lie. They need extra help so that they become secure in the earlier stages of development. It is not helpful to wait for them to become 'ready'. For example, children who are not speaking in sentences are not going to be able to write in sentences. They will need lots of stimulating experiences to help them develop their communication. That's why the time you spend listening to them and having conversations with them is so important.

Health colleagues, like health visitors or speech and language therapists, offer vital extra support to this work.

It is vital that we get to know and value all young children. All children learn more in the period from birth to five years old than any other time in their lives. If children are at risk of falling behind the majority, the best time to help them to catch up and keep up is in the early years. Every child can make progress, if they are given the right support.

When we succeed in giving every child the best start in their early years, we give them what they need today. We also set them up with every chance of success tomorrow.

Seven key features of effective practice

1. The best for every child

- All children deserve to have an equal chance of success.
- High-quality early education is good for all children. It is especially important for children from disadvantaged backgrounds.
- When they start school, children from disadvantaged backgrounds are, on average, 4 months[1] behind their peers. We need to do more to narrow that gap.
- Children who have lived through difficult experiences can begin to grow stronger when they experience high quality early education and care.
- High-quality early education and care is inclusive. Children's special educational needs and disabilities (SEND) are identified quickly. All children promptly receive any extra help they need, so they can progress well in their learning.

2. High-quality care

- The child's experience must always be central to the thinking of every practitioner.
- Babies, toddlers and young children thrive when they are loved and well cared for.
- High-quality care is consistent. Every practitioner needs to enjoy spending time with young children.
- Effective practitioners are responsive to children and babies. They notice when a baby looks towards them and gurgles and respond with pleasure.
- Practitioners understand that toddlers are learning to be independent, so they will sometimes get frustrated.
- Practitioners know that starting school, and all the other transitions in the early years, are big steps for small children.

---

[1] Education Policy Institute: Education in England Annual Report 2020 (https://epi.org.uk/wp-content/uploads/2020/09/EPI_2020_Annual_Report_.pdf) and Early years foundation stage profile results: 2018 to 2019 (https://www.gov.uk/government/statistics/early-years-foundation-stage-profile-results-2018-to-2019)

3. The curriculum: what we want children to learn

- The curriculum is a top-level plan of everything the early years setting wants the children to learn.
- Planning to help every child to develop their language is vital.
- The curriculum needs to be ambitious. Careful sequencing will help children to build their learning over time.
- Young children's learning is often driven by their interests. Plans need to be flexible.
- Babies and young children do not develop in a fixed way. Their development is like a spider's web with many strands, not a straight line.
- Depth in early learning is much more important than covering lots of things in a superficial way.

4. Pedagogy: helping children to learn

- Children are powerful learners. Every child can make progress in their learning, with the right help.
- Effective pedagogy is a mix of different approaches. Children learn through play, by adults modelling, by observing each other, and through guided learning and direct teaching.
- Practitioners carefully organise enabling environments for high-quality play. Sometimes, they make time and space available for children to invent their own play. Sometimes, they join in to sensitively support and extend children's learning.
- Children in the early years also learn through group work, when practitioners guide their learning.
- Older children need more of this guided learning.
- A well-planned learning environment, indoors and outside, is an important aspect of pedagogy.

5. Assessment: checking what children have learnt

- Assessment is about noticing what children can do and what they know. It is not about lots of data and evidence.

- Effective assessment requires practitioners to understand child development. Practitioners also need to be clear about what they want children to know and be able to do.

- Accurate assessment can highlight whether a child has a special educational need and needs extra help.

- Before assessing children, it's a good idea to think about whether the assessments will be useful.

- Assessment should not take practitioners away from the children for long periods of time.

6. Self-regulation and executive function

- Executive function includes the child's ability to:
  - hold information in mind
  - focus their attention
  - regulate their behaviour
  - plan what to do next.

- These abilities contribute to the child's growing ability to self-regulate:
  - focus their thinking
  - monitor what they are doing and adapt
  - regulate strong feelings
  - be patient for what they want
  - bounce back when things get difficult.

- Language development is central to self-regulation: children use language to guide their actions and plans. Pretend play gives many opportunities for children to focus their thinking, persist and plan ahead.

7. Partnership with parents

- It is important for parents and early years settings to have a strong and respectful partnership. This sets the scene for children to thrive in the early years.

- This includes listening regularly to parents and giving parents clear information about their children's progress.

- The help that parents give their children at home has a very significant impact on their learning.
- Some children get much less support for their learning at home than others. By knowing and understanding all the children and their families, settings can offer extra help to those who need it most.
- It is important to encourage all parents to chat, play and read with their children.

The characteristics of effective teaching and learning

In planning and guiding what children learn, practitioners must reflect on the different rates at which children are developing and adjust their practice appropriately. Three characteristics of effective teaching and learning are:

- playing and exploring - children investigate and experience things, and 'have a go'
- active learning - children concentrate and keep on trying if they encounter difficulties, and enjoy achievements
- creating and thinking critically - children have and develop their own ideas, make links between ideas, and develop strategies for doing things

Statutory framework for the early years foundation stage: early adopter version

Playing and exploring

Children will be learning to:	Examples of how to support this:
Realise that their actions have an effect on the world, so they want to keep repeating them.	Encourage babies' exploration of the world around them. Suggestions: investigating the feel of their key person's hair or reaching for a blanket in their cot. Play games like 'Peepo'. As they get more familiar, the baby or toddler will increasingly lead the play and want the adult to respond.
Reach for and accept objects. Make choices and explore different resources and materials.	Show and give babies interesting things, such as a rattle or a soft toy. Arrange for babies to take part in Treasure Basket play. Offer open-ended resources for babies and toddlers to play freely with, outdoors and inside.
Plan and think ahead about how they will explore or play with objects.	Provide different pebbles, shells and other natural materials for children to explore and arrange freely.
Guide their own thinking and actions by talking to themselves while playing. For example, a child doing a jigsaw might whisper under their breath: "Where does that one go? – I need to find the big horse next."	Help children to develop more control over their actions by giving them many opportunities to play freely and find their own ways of solving problems. When appropriate, sensitively provide a helpful commentary. You might suggest: "Why don't you look for the biggest pieces first?" That will help a child who is trying to solve a jigsaw. Children may copy your commentary by talking out

Children will be learning to:	Examples of how to support this:
	loud to themselves first. In time, this will develop into their 'inner voice'.
Make independent choices. Do things independently that they have been previously taught.	Provide a well-organised environment so that children know where materials and tools are and can access them easily. Provide enough materials and arrange spaces so that children can collaborate and learn alongside peers. Once children know how to use scissors, they can use this skill to achieve what they want to do. For example, they may want to make a mask or cut out material for a collage.
Bring their own interests and fascinations into early years settings. This helps them to develop their learning.	Extend children's interests by providing stimulating resources for them to play with, on their own and with peers, in response to their fascinations. Join in with children's play and investigations, without taking over. Talk with them about what they are doing and what they are noticing. Provide appropriate non-fiction books and links to information online to help them follow their interests.
Respond to new experiences that you bring to their attention.	Regularly provide new materials and interesting things for children to explore and investigate. Introduce children to different styles of music and art. Give them the opportunity to observe changes in living things in the setting, and around the local environment. Take children to new places, like a local theatre or museum.

Active learning

Children will be learning to:	Examples of how to support this:
Participate in routines, such as going to their cot or mat when they want to sleep. Begin to predict sequences because they know routines. For example, they may anticipate lunch when they see the table being set, or get their coat when the door to the outdoor area opens.	Help babies, toddlers and young children feel safe, secure and treasured as individuals. The key person approach gives children a secure base of care and affection, together with supportive routines. That can help them to explore and play confidently.
Show goal-directed behaviour. For example, babies may pull themselves up by using the edges of a low table to reach for a toy on top of the table. Toddlers might turn a storage box upside down so they can stand on it and reach up for an object.	Provide furniture and boxes at the right height to encourage babies to pull themselves up and reach for objects. Opportunities to play and explore freely, indoors and outside, are fun. They also help babies, toddlers and young children to develop their self-regulation as they enjoy hands-on learning and sometimes talk about what they are doing.
Use a range of strategies to reach a goal they have set themselves.	Provide plenty of high-quality, open-ended resources for children to play with freely, inside and outdoors. Suggestion: children can use wooden blocks to make lots of different structures.
Begin to correct their mistakes themselves. For example, instead of using increasing force to push a puzzle piece into the slot, they try another piece to see if it will fit.	Help young children to develop by accepting the pace of their learning. Give them plenty of time to make connections and repeat activities.
Keep on trying when things are difficult.	Help children to think about what will support them most, taking care not to offer help too soon. Some children learn by repeating something hard on their own. They learn through trial and error. Others learn by asking a friend or an adult for help. Others learn by modelling. They watch what you do or what other children do.

Creating and thinking critically

Children will be learning to	Examples of how to support this:
Take part in simple pretend play. For example, they might use an object like a brush to pretend to brush their hair, or 'drink' from a pretend cup. Sort materials. For example, at tidy-up time, children know how to put different construction materials in separate baskets.	Help babies, toddlers and young children to find their own ideas by providing open-ended resources that can be used in many different ways. Encourage and enjoy children's creative thinking as they find new ways to do things. Children need consistent routines and plenty of time so that play is not constantly interrupted. It is important to be reflective and flexible.
Review their progress as they try to achieve a goal. Check how well they are doing. Solve real problems: for example, to share nine strawberries between three friends, they might put one in front of each, then a second, and finally a third. Finally, they might check at the end that everyone has the same number of strawberries.	Help children to reflect on and talk about their learning through using photographs and learning journeys. Share in children's pride about their achievements and their enjoyment of special memories. Suggestion: you could prompt a conversation with questions like: "Do you remember when...?", "How would you do that now?" or "I wonder what you were thinking then?"
Use pretend play to think beyond the 'here and now' and to understand another perspective. For example, a child role-playing the billy goats gruff might suggest that "Maybe the troll is lonely and hungry? That's why he is fierce."	Help children to extend their ideas through sustained discussion that goes beyond what they, and you, have noticed. Consider 'how' and 'why' things happen.
Know more, so feel confident about coming up with their own ideas. Make more links between those ideas.	Help children to look come up with their own ideas and explanations. Suggestion: you could look together at woodlice outdoors with the magnifying app on a tablet. You could ask: "What's similar about woodlice and other insects?" You could use and explain terms like 'antennae' and 'thorax'.
Concentrate on achieving something that's important to them. They are increasingly able to control their attention and ignore distractions.	Offer children many different experiences and opportunities to play freely and to explore and investigate. Make time and space for children to become deeply involved in imaginative play, indoors and outside.

Communication and language

The development of children's spoken language underpins all seven areas of learning and development. Children's back-and-forth interactions from an early age form the foundations for language and cognitive development. The number and quality of the conversations they have with adults and peers throughout the day in a language-rich environment is crucial. By commenting on what children are interested in or doing, and echoing back what they say with new vocabulary added, practitioners will build children's language effectively. Reading frequently to children, and engaging them actively in stories, non-fiction, rhymes and poems, and then providing them with extensive opportunities to use and embed new words in a range of contexts, will give children the opportunity to thrive. Through conversation, story-telling and role play, where children share their ideas with support and modelling from their teacher, and sensitive questioning that invites them to elaborate, children become comfortable using a rich range of vocabulary and language structures.

Statutory framework for the early years foundation stage: early adopter version

English as an additional language

Speaking more than one language has lots of advantages for children. It is the norm in many countries around the world. Children will learn English from a strong foundation in their home language. It is important for you to encourage families to use their home language for linguistic as well as cultural reasons. Children learning English will typically go through a quiet phase when they do not say very much and may then use words in both languages in the same sentence. Talk to parents about what language they speak at home, try and learn a few key words and celebrate multilingualism in your setting.

Birth to three - babies, toddlers and young children will be learning to:	**Examples of how to support this:**
Turn towards familiar sounds. They are also startled by loud noises and accurately locate the source of a familiar person's voice, such as their key person or a parent. Gaze at faces, copying facial expressions and movements like sticking out their tongue. Make eye contact for longer periods. Watch someone's face as they talk.	Babies and toddlers thrive when you show a genuine interest in them, join in and respond warmly. Using exaggerated intonation and a sing-song voice (infant-directed speech) helps babies tune in to language. Regularly using the babies and toddlers' names helps them to pay attention to what the practitioner is saying for example: "Chloe, have some milk." It is important to minimise background noise, so don't have music playing all the time.

Birth to three - babies, toddlers and young children will be learning to:	Examples of how to support this:
Copy what adults do, taking 'turns' in conversations (through babbling) and activities. Try to copy adult speech and lip movements. Enjoy singing, music and toys that make sounds. Recognise and are calmed by a familiar and friendly voice. Listen and respond to a simple instruction.	Babies love singing and music. Sing a range of songs and play a wide range of different types of music. Move with babies to music. Babies and toddlers love action rhymes and games like 'Peepo'. As they begin to join in with the words and the actions, they are developing their attention and listening. Allow babies time to anticipate words and actions in favourite songs.
Observation checkpoint	Around 6 months, does the baby respond to familiar voices, turn to their own name and 'take turns' in conversations with babbling? Around 12 months, does the baby 'take turns' by babbling and using single words? Does the baby point to things and use gestures to show things to adults and share interests? Around 18 months, is the toddler listening and responding to a simple instruction like: "Adam, put on your shoes?"
Make sounds to get attention in different ways (for example, crying when hungry or unhappy, making gurgling sounds, laughing, cooing or babbling). Babble, using sounds like 'ba-ba', 'mamama'. Use gestures like waving and pointing to communicate.	Take time and 'tune in' to the messages babies are giving you through their vocalisations, body language and gestures. When babies and toddlers are holding and playing with objects, say what they are doing for example: "You've got the ball," and "Shake the rattle."
Reach or point to something they want while making sounds. Copy your gestures and words.	Where you can, give meaning to the baby's gestures and pointing for example: "Oh, I see, you want the teddy." Chat with babies and toddlers all the time, but be careful not to overwhelm them with talk. Allow babies

Birth to three - babies, toddlers and young children will be learning to:	Examples of how to support this:
Constantly babble and use single words during play. Use intonation, pitch and changing volume when 'talking'.	and toddlers to take the lead and then respond to their communications. Wait for the baby or toddler to speak or communicate with a sound or a look first – so that they are leading the conversation. When responding, expand on what has been said (for example, add a word). If a baby says "bottle", you could say "**milk** bottle". In a natural way, use the same word repeatedly in different contexts: "Look, a bottle of milk– oh, you've finished your bottle." Adding a word while a toddler is playing gives them the model of an expanded phrase. It also keeps the conversation on their topic of interest. Suggestion: if they say "bag", you could say: "Yes, daddy's bag".
Observation checkpoint	Is the baby using speech sounds (babbling) to communicate with adults? Around 12 months: is the baby beginning to use single words like mummum, dada, tete (teddy)? Around 15 months, can the baby say around 10 words (they may not all be clear)? Around 18 months, is the toddler using a range of adult like speech patterns (jargon) and at least 20 clear words?
Understand single words in context – 'cup', 'milk', 'daddy'. Understand frequently used words such as 'all gone', 'no' and 'bye-bye'.	You can help babies with their understanding by using gestures and context. Suggestion: point to the cup and say "cup". Talking about what you are doing helps babies learn language in context. Suggestion: "I'm pouring out your milk into the cup".
Observation checkpoint	Around 12 months, can the baby choose between 2 objects: "Do you want the ball or the car?"
Understand simple instructions like "give to nanny" or "stop". Recognise and point to objects if asked about them.	Singing, action rhymes and sharing books give children rich opportunities to understand new words. Play with groups of objects (different small world animals, or soft toys, or tea and picnic sets). Make sure you name things whilst playing, and talk about what you are doing

Birth to three - babies, toddlers and young children will be learning to:	Examples of how to support this:
Observation checkpoint	Around 18 months, does the toddler understand lots of different single words and some two-word phrases, such as "give me" or "shoes on"?
Generally focus on an activity of their own choice and find it difficult to be directed by an adult.	Help toddlers and young children to focus their attention by using their name: "Fatima, put your coat on".
Listen to other people's talk with interest, but can easily be distracted by other things.	You can help toddlers and young children listen and pay attention by using gestures like pointing and facial expressions.
Make themselves understood, and can become frustrated when they can't. Start to say how they are feeling, using words as well as actions.	You can help toddlers who are having tantrums by being calm and reassuring. Help toddlers to express what's angering them by suggesting words to describe their emotions, like 'sad' or 'angry'. You can help further by explaining in simple terms why you think they may be feeling that emotion.
Start to develop conversation, often jumping from topic to topic. Develop pretend play: 'putting the baby to sleep' or 'driving the car to the shops'.	Make time to connect with babies, toddlers and young children. Tune in and listen to them and join in with their play, indoors and outside. Allow plenty of time to have conversations together, rather than busily rushing from one activity to the next. When you know a young child well, it is easier to understand them and talk about what their family life. For example: "OK, I see. You went to the shops with Aunty Maya".
Observation checkpoint	By around 2 years old, is the child showing an interest in what other children are playing and sometimes joins in? By around 3 years old, can the child shift from one task to another if you get their attention. Using the child's name can help: "Jason, can you stop now? We're tidying up".
Use the speech sounds p, b, m, w. Pronounce: - l/r/w/y - f/th - s/sh/ch/dz/j	Toddlers and young children will pronounce some words incorrectly. Instead of correcting them, reply to what they say and use the words they have mispronounced. Children will then learn from your positive model, without losing the confidence to speak. Toddlers and young children sometimes hesitate and repeat sounds and words when thinking what to say.

Birth to three - babies, toddlers and young children will be learning to:	Examples of how to support this:
- multi-syllabic words such as 'banana' and 'computer'	Listen patiently. Do not say the words for them. If the child or parents are distressed or worried by this, contact a speech and language therapist for advice. Encourage children to talk. Do not use too many questions: four comments to every question is a useful guide.
Observation checkpoint	Towards their second birthday, can the child use up to 50 words? Is the child beginning to put two or three words together: "more milk"? Is the child frequently asking questions, such as the names of people and objects? Towards their third birthday, can the child use around 300 words? These words include descriptive language. They include words for time (for example, 'now' and 'later'), space (for example, 'over there') and function (for example, they can tell you a sponge is for washing). Is the child linking up to 5 words together? Is the child using pronouns ('me', 'him', 'she'), and using plurals and prepositions ('in', 'on', 'under') - these may not always be used correctly to start with. Can the child follow instructions with three key words like: "Can you **wash dolly's face**?"
Listen to simple stories and understand what is happening, with the help of the pictures.	Share picture books every day with children. Encourage them to talk about the pictures and the story. Comment on the pictures – for example: "It looks like the boy is a bit worried..." and wait for their response. You might also ask them about the pictures: "I wonder what the caterpillar is doing now?" Books with just pictures and no words can especially encourage conversations. Tell children the names of things they do not know and choose books that introduce interesting new vocabulary to them.

Birth to three - babies, toddlers and young children will be learning to:	**Examples of how to support this:**
Identify familiar objects and properties for practitioners when they are described: for example: 'Katie's coat', 'blue car', 'shiny apple'. Understand and act on longer sentences like 'make teddy jump' or 'find your coat'.	When appropriate, you can check children's understanding by asking them to point to particular pictures. Or ask them to point to particular objects in a picture. For example: "Can you show me the big boat?"
Understand simple questions about 'who', 'what' and 'where' (but generally not 'why').	When talking with young children, give them plenty of processing time (at least 10 seconds). This gives them time to understand what you have said and think of their reply.
Observation checkpoint	Around the age of 2, can the child understand many more words than they can say – between 200–500 words? Around the age of 2, can the child understand simple questions and instructions like: "Where's your hat?" or "What's the boy in the picture doing?" Around the age of 3, can the child show that they understand action words by pointing to the right picture in a book. For example: "Who's jumping?" Note: watch out for children whose speech is not easily understood by unfamiliar adults. Monitor their progress and consider whether a hearing test might be needed.

3 & 4-year-olds will be learning to:	**Examples of how to support this:**
Enjoy listening to longer stories and can remember much of what happens. Pay attention to more than one thing at a time, which can be difficult.	Offer children at least a daily story time as well as sharing books throughout the session. If they are busy in their play, children may not be able to switch their attention and listen to what you say. When you need to, help young children to switch their attention from what they are doing to what you are saying. Give them a clear prompt. Suggestion: say the child's name and then: "Please stop and listen".

3 & 4-year-olds will be learning to:	Examples of how to support this:
Use a wider range of vocabulary. Understand a question or instruction that has two parts, such as: "Get your coat and wait at the door". Understand 'why' questions, like: "Why do you think the caterpillar got so fat?"	Extend children's vocabulary, explaining unfamiliar words and concepts and making sure children have understood what they mean through stories and other activities. These should include words and concepts which occur frequently in books and other contexts, but are not used every day by many young children. Suggestion: use scientific vocabulary when talking about the parts of a flower or an insect, or different types of rocks. Examples from 'The Gruffalo' include: 'stroll', 'roasted', 'knobbly', 'wart' and 'feast'. Provide children with a rich language environment by sharing books and activities with them. Encourage children to talk about what is happening and give their own ideas. High-quality picture books are a rich source for learning new vocabulary and more complex forms of language: "Excuse me, I'm very hungry. Do you think I could have tea with you?" Shared book-reading is a powerful way of having extended conversations with children. It helps children to build their vocabulary. Offer children lots of interesting things to investigate, like different living things. This will encourage them to ask questions.
Sing a large repertoire of songs. Know many rhymes, be able to talk about familiar books, and be able to tell a long story.	Consider which core books, songs and rhymes you want children to become familiar with and grow to love. Activities planned around those core books will help the children to practise the vocabulary and language from those books. It will also support their creativity and play. Small world play based on 'Dear Zoo' will help children to learn the names of the different animals. Or they could shop for the different types of fruit in 'Handa's Surprise'. Pick them out and talk about how they look. This will help children to name the different types of fruit. Back in the setting, taste them and talk about their texture and smell.

3 & 4-year-olds will be learning to:	Examples of how to support this:
	Outdoor play themed around 'We're Going a Bear Hunt' might lead to the children creating their own 'hunts' and inventing their own rhymes.
Develop their communication, but may continue to have problems with irregular tenses and plurals, such as 'runned' for 'ran', 'swimmed' for 'swam'. Develop their pronunciation but may have problems saying: - some sounds: r, j, th, ch, and sh - multi-syllabic words such as 'pterodactyl', 'planetarium' or 'hippopotamus'.	Children may use ungrammatical forms like 'I swimmed'. Instead of correcting them, recast what the child said. For example: "How lovely that **you swam** in the sea on holiday". When children have difficulties with correct pronunciation, reply naturally to what they say. Pronounce the word correctly so they hear the correct model.
Use longer sentences of four to six words.	Expand on children's phrases. For example, if a child says, "going out shop", you could reply: "Yes, Jason is going to the shop". As well as adding language, add new ideas. For example: "I wonder if they'll get the 26 bus?"
Be able to express a point of view and to debate when they disagree with an adult or a friend, using words as well as actions. Start a conversation with an adult or a friend and continue it for many turns. Use talk to organise themselves and their play: "Let's go on a bus... you sit there... I'll be the driver."	Model language that promotes thinking and challenges children: "I can see that's empty – I wonder what happened to the snail that used to be in that shell?" Open-ended questions like "I wonder what would happen if….?" encourage more thinking and longer responses. Sustained shared thinking is especially powerful. This is when two or more individuals (adult and child, or children) 'work together' in an intellectual way to solve a problem, clarify a concept, evaluate activities, extend a narrative etc. Help children to elaborate on how they are feeling: "You look sad. Are you upset because Jasmin doesn't want to do the same thing as you?"

3 & 4-year-olds will be learning to:	Examples of how to support this:
Observation checkpoint	Around the age of 3, can the child shift from one task to another if you fully obtain their attention, for example, by using their name? Around the age of 4, is the child using sentences of four to six words – “I want to play with cars” or “What’s that thing called?”? Can the child use sentences joined up with words like ‘because’, ‘or’, ‘and’? For example: “I like ice cream because it makes my tongue shiver”. Is the child using the future and past tense: “I am going to the park” and “I went to the shop”? Can the child answer simple ‘why’ questions?

Children in reception will be learning to:	Examples of how to support this:
Understand how to listen carefully and why listening is important.	Promote and model active listening skills: “Wait a minute, I need to get into a good position for listening, I can’t see you. Let’s be quiet so I can concentrate on what you’re saying.” Signal when you want children to listen: “Listen carefully now for how many animals are on the broom.” Link listening with learning: “I could tell you were going to say the right answer, you were listening so carefully.”
Learn new vocabulary.	Identify new vocabulary before planning activities, for example, changes in materials: ‘dissolving’, ‘drying’, ‘evaporating’; in music: ‘percussion’, ‘tambourine’. Bring in objects, pictures and photographs to talk about, for example vegetables to taste, smell and feel.

Children in reception will be learning to:	Examples of how to support this:
	Discuss which category the word is in, for example: "A cabbage is a kind of vegetable. It's a bit like a sprout but much bigger". Have fun saying the word in an exaggerated manner. Use picture cue cards to talk about an object: "What colour is it? Where would you find it? What shape is it? What does it smell like? What does it look like? What does it feel like? What does it sound like? What does it taste like?"
Use new vocabulary through the day.	Model words and phrases relevant to the area being taught, deliberately and systematically: "I'm thrilled that everyone's on time today", "I can see that you're delighted with your new trainers", "Stop shrieking, you're hurting my ears!", "What a downpour – I've never seen so much rain!", "It looks as if the sun has caused the puddles to evaporate", "Have you ever heard such a booming voice?" Use the vocabulary repeatedly through the week. Keep a list of previously taught vocabulary and review it in different contexts.
Ask questions to find out more and to check they understand what has been said to them.	Show genuine interest in knowing more: "This looks amazing, I need to know more about this." Think out loud, ask questions to check your understanding; make sure children can answer who, where and when questions before you move on to why and 'how do you know' questions: "I wonder why this jellyfish is so dangerous? Ahh, it has poison in its tentacles."
Articulate their ideas and thoughts in well-formed sentences.	Use complete sentences in your everyday talk. Help children build sentences using new vocabulary by rephrasing what they say and structuring their responses using sentence starters. Narrate your own and children's actions: "I've never seen so many beautiful bubbles, I can see all the colours of the rainbow in them."

Children in reception will be learning to:	Examples of how to support this:
	Build upon their incidental talk: "Your tower is definitely the tallest I've seen all week. Do you think you'll make it any higher?" Suggestion: ask open questions - "How did you make that? Why does the wheel move so easily? What will happen if you do that?" Instead of correcting, model accurate irregular grammar such as past tense, plurals, complex sentences: "That's right: you drank your milk quickly; you were quicker than Darren."
Connect one idea or action to another using a range of connectives.	Narrate events and actions: "I knew it must be cold outside because he was putting on his coat and hat." Remind children of previous events: "Do you remember when we forgot to wear our raincoats last week? It poured so much that we got drenched!" Extend their thinking: "You've thought really hard about building your tower, but how will you stop it falling down?"
Describe events in some detail.	Make deliberate mistakes highlighting to children that sometimes you might get it wrong: "It's important to get things in the right order so that people know what I'm talking about. Listen carefully to see if I have things in the right order: 'last week..." Use sequencing words with emphasis in your own stories: "Before school I had a lovely big breakfast, then I had a biscuit at break time and after that I had two pieces of fruit after lunch. I'm so full!"
Use talk to help work out problems and organise thinking and activities, and to explain how things work and why they might happen.	Think out loud how to work things out. Encourage children to talk about a problem together and come up with ideas for how to solve it. Give children problem solving words and phrases to use in their explanations: 'so that', 'because', 'I think it's...', 'you could...', 'it might be...'
Develop social phrases.	Model talk routines through the day. For example, arriving in school: "Good morning, how are you?"
Engage in storytimes.	Timetable a storytime at least once a day.

Children in reception will be learning to:	Examples of how to support this:
	Draw up a list of books that you enjoy reading aloud to children, including traditional and modern stories. Choose books that will develop their vocabulary. Display quality books in attractive book corners. Send home familiar and good-quality books for parents to read aloud and talk about with their children. Show parents how to share stories with their children.
Listen to and talk about stories to build familiarity and understanding.	Read and re-read selected stories. Show enjoyment of the story using your voice and manner to make the meaning clear. Use different voices for the narrator and each character. Make asides, commenting on what is happening in a story: "That looks dangerous – I'm sure they're all going to fall off that broom!" Link events in a story to your own experiences. Talk about the plot and the main problem in the story. Identify the main characters in the story, and talk about their feelings, actions and motives. Take on different roles in imaginative play, to interact and negotiate with people in longer conversations. Practise possible conversations between characters.
Retell the story, once they have developed a deep familiarity with the text; some as exact repetition and some in their own words.	Make familiar books available for children to share at school and at home. Make time for children to tell each other stories they have heard, or to visitors.
Use new vocabulary in different contexts.	Have fun with phrases from the story through the day: "I searched for a pencil, but no pencil could be found."

Children in reception will be learning to:	Examples of how to support this:
	Explain new vocabulary in the context of story, rather than in word lists.
Listen carefully to rhymes and songs, paying attention to how they sound.	Show your enjoyment of poems using your voice and manner to give emphasis to carefully chosen words and phrases. Model noticing how some words sound: "That poem was about a frog on a log; those words sound a bit the same at the end don't they? They rhyme." In poems and rhymes with very regular rhythm patterns, pause before the rhyming word to allow children to join in or predict the word coming next. Encourage children to have fun with rhyme, even if their suggestions don't make complete sense. Choose a few interesting longer words from the poem, rhyme or song and clap out their beat structure, helping children to join in with the correct number of 'claps'.
Learn rhymes, poems and songs.	Select traditional and contemporary poems and rhymes to read aloud to children. Help children to join in with refrains and learn some verses by heart using call and response. When singing songs by heart, talk about words in repeated phrases from within a refrain or verse so that word boundaries are noticed and not blurred: "Listen carefully, what words can you hear? Oncesuppona time: once – upon – a – time."
Engage in non-fiction books.	Read aloud books to children that will extend their knowledge of the world and illustrate a current topic. Select books containing photographs and pictures, for example, places in different weather conditions and seasons.
Listen to and talk about selected non-fiction to develop a deep familiarity with new knowledge and vocabulary.	Re-read some books so children learn the language necessary to talk about what is happening in each illustration and relate it to their own lives. Make the books available for children to share at school and at home.

Personal, Social and Emotional Development

Children's personal, social and emotional development (PSED) is crucial for children to lead healthy and happy lives, and is fundamental to their cognitive development. Underpinning their personal development are the important attachments that shape their social world. Strong, warm and supportive relationships with adults enable children to learn how to understand their own feelings and those of others. Children should be supported to manage emotions, develop a positive sense of self, set themselves simple goals, have confidence in their own abilities, to persist and wait for what they want and direct attention as necessary. Through adult modelling and guidance, they will learn how to look after their bodies, including healthy eating, and manage personal needs independently. Through supported interaction with other children they learn how to make good friendships, co-operate and resolve conflicts peaceably. These attributes will provide a secure platform from which children can achieve at school and in later life.

Statutory framework for the early years foundation stage: early adopter version

Birth to three - babies, toddlers and young children will be learning to:	Examples of how to support this:
Find ways to calm themselves, through being calmed and comforted by their key person.	When settling a baby or toddler into nursery, the top priority is for the key person to develop a strong and loving relationship with the young child. Learn from the family about what they do to soothe their child and what to look out for – for example, a baby who scratches at their head when they are getting tired. Find out what calms a baby – rocking, cuddling or singing. Make sure babies and toddlers can get hold of their comfort object when they need it. Explain to parents that once babies establish 'object permanence', they become more aware of the presence or absence of their parents. Object permanence means knowing that something continues to exist even when out of sight. This can make separations much more distressing and difficult between 6–24 months.

Birth to three - babies, toddlers and young children will be learning to:	Examples of how to support this:
Establish their sense of self.	Babies develop a sense of self by interacting with others, and by exploring their bodies and objects around them, inside and outdoors. Respond and build on babies' expressions and gestures, playfully exploring the idea of self/other. Suggestion: point to your own nose/eyes/mouth, point to the baby's.
Express preferences and decisions. They also try new things and start establishing their autonomy. Engage with others through gestures, gaze and talk. Use that engagement to achieve a goal. For example, gesture towards their cup to say they want a drink.	Be positive and interested in what babies do as they develop their confidence in trying new things. Help toddlers and young children to make informed choices from a limited range of options. Suggestion: enable children to choose which song to sing from a set of four song cards, by pointing. Enable children to choose whether they want milk or water at snack time.
Find ways of managing transitions, for example from their parent to their key person.	Support children as they find their own different ways to manage feelings of sadness when their parents leave them. Some children might need to hold onto a special object from home to feel strong and confident in the setting. Some might need to snuggle in and be comforted by their key person. Some might get busy straight away in their favourite play or with another child they feel close to. Young children need to feel secure as they manage difficult emotions. Provide consistent and predictable routines, with flexibility when needed.
Thrive as they develop self-assurance.	Provide consistent, warm and responsive care. At first, centre this on the key person. In time, children can develop positive relationships with other adults. When the key person is not available, make sure that someone familiar provides comfort and support, and carries out intimate care routines.

Birth to three - babies, toddlers and young children will be learning to:	Examples of how to support this:
Look back as they crawl or walk away from their key person. Look for clues about how to respond to something interesting. Play with increasing confidence on their own and with other children, because they know their key person is nearby and available. Feel confident when taken out around the local neighbourhood, and enjoy exploring new places with their key person.	Acknowledge babies' and toddlers' brief need for reassurance as they move away from their key person. Encourage babies and toddlers to explore, indoors and outside. Help them to become more independent by smiling and looking encouraging, for example when a baby keeps crawling towards a rattle. Arrange resources inside and outdoors to encourage children's independence and growing self-confidence. Suggestion: Treasure Basket play allows babies who can sit up to choose what to play with. Store resources so that children can access them freely, without needing help.
Feel strong enough to express a range of emotions. Grow in independence, rejecting help ("me do it"). Sometimes this leads to feelings of frustration and tantrums.	Help children to feel emotionally safe with a key person and, gradually, with other members of staff. Show warmth and affection, combined with clear and appropriate boundaries and routines. Develop a spirit of friendly co-operation amongst children and adults. Encourage children to express their feelings through words like 'sad', 'upset' or 'angry'. Toddlers and young children may have periods of time when their favourite word is 'no' and when they want to carry out their wishes straight away. Maintain sensible routines and boundaries for children during these testing times. Negative or harsh responses can cause children to feel unduly anxious and emotionally vulnerable. Offer supervision or work discussion sessions to staff. Staff will need to talk about the strong feelings that children may express. How are practitioners feeling about these and developing their understanding of the children's feelings?

Birth to three - babies, toddlers and young children will be learning to:	Examples of how to support this:
Begin to show 'effortful control'. For example, waiting for a turn and resisting the strong impulse to grab what they want or push their way to the front. Be increasingly able to talk about and manage their emotions.	When appropriate, notice and talk about children's feelings. For example: "I can see it's hard to wait, just a minute and then it's your turn to go down the slide." Model useful phrases like "Can I have a turn?" or "My turn next."
Notice and ask questions about differences, such as skin colour, types of hair, gender, special needs and disabilities, and so on.	Be open to what children say about differences and answer their questions straightforwardly. Help children develop positive attitudes towards diversity and inclusion. Help all children to feel that they are valued, and they belong.
Develop friendships with other children.	Support children to find ways into the play and friendship groups of others. For example, encourage them to stand and watch from the side with you. Talk about what you see, and suggest ways for the child to join in.
Safely explore emotions beyond their normal range through play and stories. Talk about their feelings in more elaborated ways: "I'm sad because..." or "I love it when ...".	Story times with props can engage children in a range of emotions. They can feel the family's fear as the bear chases them at the end of 'We're Going on a Bear Hunt'. They can feel relief when the Gruffalo is scared away by the mouse. Recognise, talk about and expand on children's emotions. For example, you might say: "Sara is smiling. She really wanted a turn with the truck."
Observation checkpoint	Around 7 months, does the baby respond to their name and respond to the emotions in your voice? Around 12 months, does the baby start to be shy around strangers and show preferences for certain people and toys?

Birth to three - babies, toddlers and young children will be learning to:	Examples of how to support this:
	Around 18 months, is the toddler increasingly curious about their world and wanting to explore it and be noticed by you? Around the age of 2, does the child start to see themselves as a separate person? For example, do they decide what to play with, what to eat, what to wear? Between the ages of 2 and 3, does the child start to enjoy the company of other children and want to play with them? Note: watch out for children who get extremely upset by certain sounds, smells or tastes, and cannot be calmed. Or children who seem worried, sad or angry for much of the time. You will need to work closely with parents and other agencies to find out more about these developmental difficulties.

3 & 4-year-olds will be learning to:	Examples of how to support this:
Select and use activities and resources, with help when needed. This helps them to achieve a goal they have chosen, or one which is suggested to them.	Respond to children's increasing independence and sense of responsibility. As the year proceeds, increase the range of resources and challenges, outdoors and inside. One example of this might be starting the year with light hammers, plastic golf tees and playdough. This equipment will offer children a safe experience of hammering. Wait until the children are ready to follow instructions and use tools safely. Then you could introduce hammers with short handles, nails with large heads, and soft blocks of wood. Widen the range of activities that children feel confident to take part in, outdoors and inside. Model inviting new activities that encourage children to come over and join in, such as folding paper to make animals, sewing or weaving.
Develop their sense of responsibility and membership of a community.	Give children appropriate tasks to carry out. Suggestion: they can fetch milk cartons or fruit. They can wash up their own plates after their snack.

3 & 4-year-olds will be learning to:	Examples of how to support this:
Become more outgoing with unfamiliar people, in the safe context of their setting. Show more confidence in new social situations.	Invite trusted people into the setting to talk about and show the work they do. Some examples of this might be plumbers, artists or firefighters. Take children out on short walks around the neighbourhood. When ready, take them on trips to interesting places like a local museum, theatre or place of worship.
Play with one or more other children, extending and elaborating play ideas. Find solutions to conflicts and rivalries. For example, accepting that not everyone can be Spider-Man in the game, and suggesting other ideas.	Involve children in making decisions about room layout and resources. Suggestion: you could set up a special role-play area in response to children's fascination with space. Support children to carry out decisions, respecting the wishes of the rest of the group. Further resource and enrich children's play, based on their interests. Suggestion: children often like to talk about their trips to hairdressers and barbers. You could provide wigs reflecting different ethnicities, combs and brushes etc. to stimulate pretend play around their interest. Notice children who find it difficult to play. They may need extra help to share and manage conflicts. You could set up play opportunities in quiet spaces for them, with just one or two other children. You may need to model positive play and co-operation. Teach children ways of solving conflicts. Suggestion: model how to listen to someone else and agree a compromise.
Increasingly follow rules, understanding why they are important. Remember rules without needing an adult to remind them	Explain why we have rules and display a small number of necessary rules visually as reminders. Suggestion: display a photo showing a child taking just one piece of fruit at the snack table.
Develop appropriate ways of being assertive. Talk with others to solve conflicts.	Children with high levels of negative emotion need clear boundaries and routines. They also need practitioners to interact calmly and sensitively with them. Model ways that you calm yourself down, such as stopping and taking a few deep breaths. This can help children to learning ways to calm themselves. If

3 & 4-year-olds will be learning to:	Examples of how to support this:
Talk about their feelings using words like 'happy', 'sad', 'angry' or 'worried'.	adults are excessively challenging or controlling, children can become more aggressive in the group. They may increasingly 'act out' their feelings. For example, when they feel sad, they might hit another child to make that child feel sad as well.
Understand gradually how others might be feeling.	Help children explore situations from different points of view. Talk together about how others might be feeling. Bring these ideas into children's pretend play: "I wonder how the chicken is feeling, now the fox is creeping up on her?"
Observation checkpoint	Around the age of 3, can the child sometimes manage to share or take turns with others, with adult guidance and understanding 'yours' and 'mine'? Can the child settle to some activities for a while? Around the age of 4, does the child play alongside others or do they always want to play alone? Does the child take part in pretend play (for example, being 'mummy' or 'daddy'?) Does the child take part in other pretend play with different roles – being the Gruffalo, for example? Can the child generally negotiate solutions to conflicts in their play? Note: watch out for children who seem worried, sad or angry for much of the time, children who seem to flit from one thing to the next or children who seem to stay for over-long periods doing the same thing, and become distressed if they are encouraged to do something different. You will need to work closely with parents and other agencies to find out more about these developmental difficulties.

Children in reception will be learning to:	Examples of how to support this:
See themselves as a valuable individual.	Make time to get to know the child and their family. Ask parents about the child's history, likes, dislikes, family members and culture. Take opportunities in class to highlight a child's interests, showing you know them and about them.

Build constructive and respectful relationships.	Make sure children are encouraged to listen to each other as well as the staff. Ensure children's play regularly involves sharing and cooperating with friends and other peers. Congratulate children for their kindness to others and express your approval when they help, listen and support each other. Allow children time in friendship groups as well as other groupings. Have high expectations for children following instructions, with high levels of support when necessary.
Express their feelings and consider the feelings of others.	Model positive behaviour and highlight exemplary behaviour of children in class, narrating what was kind and considerate about the behaviour. Encourage children to express their feelings if they feel hurt or upset using descriptive vocabulary. Help and reassure them when they are distressed, upset or confused. Undertake specific activities that encourage talk about feelings and their opinions.
Show resilience and perseverance in the face of challenge.	Offer constructive support and recognition of child's personal achievements. Provide opportunities for children to tell each other about their work and play. Help them reflect and self-evaluate their own work. Help them to develop problem-solving skills by talking through how they, you and others resolved a problem or difficulty. Show that mistakes are an important part of learning and going back is trial and error not failure. Help children to set own goals and to achieve them.
Identify and moderate their own feelings socially and emotionally.	Give children strategies for staying calm in the face of frustration. Talk them through why we take turns, wait politely, tidy up after ourselves and so on.

	Encourage them to think about their own feelings and those of others by giving explicit examples of how others might feel in particular scenarios. Give children space to calm down and return to an activity. Support all children to recognise when their behaviour was not in accordance with the rules and why it is important to respect class rules and behave correctly towards others.
Think about the perspectives of others.	Use dialogic story time (talking about the ideas arising from the story whilst reading aloud) to discuss books that deal with challenges, explaining how the different characters feel about these challenges and overcome them. Ask children to explain to others how they thought about a problem or an emotion and how they dealt with it.
Manage their own needs.	Model practices that support good hygiene, such as insisting on washing hands before snack time. Narrate your own decisions about healthy foods, highlighting the importance of eating plenty of fruits and vegetables.

Physical Development

Physical activity is vital in children's all-round development, enabling them to pursue happy, healthy and active lives. Gross and fine motor experiences develop incrementally throughout early childhood, starting with sensory explorations and the development of a child's strength, co-ordination and positional awareness through tummy time, crawling and play movement with both objects and adults. By creating games and providing opportunities for play both indoors and outdoors, adults can support children to develop their core strength, stability, balance, spatial awareness, co-ordination and agility. Gross motor skills provide the foundation for developing healthy bodies and social and emotional well-being. Fine motor control and precision helps with hand-eye co-ordination which is later linked to early literacy. Repeated and varied opportunities to explore and play with small world activities, puzzles, arts and crafts and the practice of using small tools, with feedback and support from adults, allow children to develop proficiency, control and confidence.

Statutory framework for the early years foundation stage: early adopter version

Birth to three - babies, toddlers and young children will be learning to:	**Examples of how to support this:**
Lift their head while lying on their front. Push their chest up with straight arms. Roll over: from front to back, then back to front. Enjoy moving when outdoors and inside.	Some babies need constant physical contact, attention and physical intimacy. Respond warmly and patiently to them. Provide adequate, clean floor space for babies to experience tummy-time and back time. Offer this frequently throughout the day so that they can develop their gross motor skills (kicking, waving, rolling and reaching).
Sit without support. Begin to crawl in different ways and directions. Pull themselves upright and bouncing in preparation for walking.	Encourage babies to sit on you, climb over you, and rock, bounce or sway with you. Notice, cherish and applaud the physical achievements of babies and toddlers. Give babies time to move freely during care routines, like nappy-changing. Encourage independence. Suggestion: offer a range of opportunities for children to move by themselves,

Birth to three - babies, toddlers and young children will be learning to:	Examples of how to support this:
	making their own decisions about direction and speed.
Reach out for objects as co-ordination develops. Eat finger food and develop likes and dislikes. Try a wider range of foods with different tastes and textures. Lift objects up to suck them. Pass things from one hand to the other. Let go of things and hand them to another person, or drop them.	Gradually share control of the bottle with young babies. Introduce children regularly and repeatedly to new foods, being positive and patient as they try new things. Value the choices children make, whilst also sensitively encouraging them to try healthy foods. Consider introducing a supervised toothbrushing programme. Use everyday, open-ended materials to support overall co-ordination. Suggestions: sponges and cloths to hold, squash and throw, or wet and squeeze. Provide a range of surfaces and materials for babies to explore, stimulating touch and all the senses.
Observation checkpoint	Does the baby move with ease and enjoyment? At around 12 months, can the baby pull to stand from a sitting position and sit down? Can the baby pick up something small with their first finger and thumb (such as a piece of string)? Note: look out for babies and young toddlers who appear underweight, overweight or to have poor dental health. You will need to work closely with parents and health visitors to help improve the child's health.
Gradually gain control of their whole body through continual practice of large movements, such as waving, kicking, rolling, crawling and walking. Clap and stamp to music.	Provide a wide range of opportunities for children to move throughout the day: indoors and outside, alone or with others, with and without apparatus. Include risky and rough and tumble play, as appropriate. Join in with children's movement play when invited and if it is appropriate. Then you can show different ways of moving and engaging with the resources.

Birth to three - babies, toddlers and young children will be learning to:	**Examples of how to support this:**
Fit themselves into spaces, like tunnels, dens and large boxes, and move around in them. Enjoy starting to kick, throw and catch balls. Build independently with a range of appropriate resources.	Help young children learn what physical risks they are confident and able to take. Encourage children to climb unaided and to stop if they do not feel safe. If you lift them onto the apparatus and hold them so they balance, they will not develop a sense of what they can do safely. Offer outdoor play every day for at least 45 minutes. Include lots of opportunities for children to move freely and explore their surroundings like a slope, a large hole, puddles or a sandpit. Consider wider opportunities for movement. Suggestions: using large moveable resources like hollow blocks, swinging on monkey bars, soft play, climbing walls, crawling into tunnels and dens. Consider going to suitable local facilities.
Begin to walk independently – choosing appropriate props to support at first. Walk, run, jump and climb – and start to use the stairs independently.	As soon as children are able, encourage 'active travel' to and from the setting – for example, walking, scooter or bike.
Spin, roll and independently use ropes and swings (for example, tyre swings). Sit on a push-along wheeled toy, use a scooter or ride a tricycle.	Provide materials and equipment that support physical development - both large and small motor skills. Encourage children to use materials flexibly and combine them in different ways. Check that children's clothing and footwear are not too tight or too large.
Observation checkpoint	Around their second birthday, can the toddler run well, kick a ball, and jump with both feet off the ground at the same time? Around their third birthday, can the child climb confidently, catch a large ball and pedal a tricycle?
Develop manipulation and control. Explore different materials and tools.	Provide different types of paper for children to tear, make marks on and print on. Provide lots of different things for young children to grasp, hold and explore, like clay, finger paint, spoons, brushes, shells.

Birth to three - babies, toddlers and young children will be learning to:	**Examples of how to support this:**
Use large and small motor skills to do things independently, for example manage buttons and zips, and pour drinks. Show an increasing desire to be independent, such as wanting to feed themselves and dress or undress.	Provide babies and toddlers with lots of opportunities to feed themselves. Encourage them to dress and undress independently. Be patient, do not rush and take time to talk about what they are doing and why: "It's a bit cold and wet today – what do we need to wear to keep warm and dry?" At meal and snack times, encourage children to try a range of foods as they become more independent eaters. Encourage children to help with carrying, pouring drinks, cleaning and sorting. Encourage young children's personal decision-making by offering real choices – water or milk, for example. They can comment on how to eat healthily, listen to children's responses and develop conversations about this. Encourage good eating habits and behaviours, such as not snatching, sharing and waiting for a second helping.
Observation checkpoint	Look out for children who find it difficult to sit comfortably on chairs. They may need help to develop their core muscles. You can help them by encouraging them to scoot on sit-down trikes without pedals, and jump on soft-play equipment.
Learn to use the toilet with help, and then independently.	You cannot force a child to use the potty or toilet. You need to establish friendly co-operation with the child. That will help them take this important step. Children can generally control their bowels before their bladder. Notice when young children are ready to begin toilet training and discuss this with their parents: - they know when they have got a wet or dirty nappy - they get to know when they are peeing and may tell you they are doing it - the gap between wetting is at least an hour - they show they need to pee by fidgeting or going somewhere quiet or hidden - they know when they need to pee and may say so in advance

Birth to three - babies, toddlers and young children will be learning to:	**Examples of how to support this:**
	Potty training is fastest if you start it when the child is at the last stage. By the age of 3, 9 out of 10 children are dry most days. All children will have the occasional 'accident', though, especially when excited, busy or upset.

3 & 4-year-olds will be learning to:	**Examples of how to support this:**
Continue to develop their movement, balancing, riding (scooters, trikes and bikes) and ball skills. Go up steps and stairs, or climb up apparatus, using alternate feet. Skip, hop, stand on one leg and hold a pose for a game like musical statues. Use large-muscle movements to wave flags and streamers, paint and make marks.	Encourage children to transfer physical skills learnt in one context to another one. Suggestion: children might first learn to hammer in pegs to mark their Forest school boundary, using a mallet. Then, they are ready to learn how to use hammers and nails at the woodwork bench. Encourage children to paint, chalk or make marks with water on large vertical surfaces. Suggestion: use walls as well as easels to stimulate large shoulder and arm movements. These experiences help children to 'cross the mid-line' of their bodies. When they draw a single line from left to right, say, they don't need to pass the paintbrush from one hand to another or have to move their whole body along.
Start taking part in some group activities which they make up for themselves, or in teams. Increasingly be able to use and remember sequences and patterns of movements which are related to music and rhythm.	Lead movement-play activities when appropriate. These will challenge and enhance children's physical skills and development – using both fixed and flexible resources, indoors and outside. Model the vocabulary of movement – 'gallop', 'slither' – and encourage children to use it. Also model the vocabulary of instruction – 'follow', 'lead', 'copy' – and encourage children to use it.
Match their developing physical skills to tasks and activities in the setting. For example, they decide whether to crawl, walk or run across a	Encourage children to become more confident, competent, creative and adaptive movers. Then, extend their learning by providing opportunities to play outdoors in larger areas, such as larger parks and spaces in the local area, or through Forest or Beach school.

3 & 4-year-olds will be learning to:	Examples of how to support this:
plank, depending on its length and width.	
Choose the right resources to carry out their own plan. For example, choosing a spade to enlarge a small hole they dug with a trowel. Collaborate with others to manage large items, such as moving a long plank safely, carrying large hollow blocks.	Explain why safety is an important factor in handling tools, and moving equipment and materials. Have clear and sensible rules for everybody to follow.
Observation checkpoint	Look out for children who appear to be overweight or to have poor dental health, where this has not been picked up and acted on at an earlier health check. Discuss this sensitively with parents and involve the child's health visitor. Adapt activities to suit their particular needs, so all children feel confident to move and take part in physical play.
Use one-handed tools and equipment, for example, making snips in paper with scissors. Use a comfortable grip with good control when holding pens and pencils. Start eating independently and learning how to use a knife and fork. Show a preference for a dominant hand.	You can begin by showing children how to use one-handed tools (scissors and hammers, for example) and then guide them with hand-over-hand help. Gradually reduce the help you are giving and allow the child to use the tool independently. The tripod grip is a comfortable way to hold a pencil or pen. It gives the child good control. The pen is pinched between the ball of the thumb and the fore-finger, supported by the middle finger with the other fingers tucked into the hand. You can help children to develop this grip with specially designed pens and pencils, or grippers. Encourage children to pick up small objects like individual gravel stones or tiny bits of chalk to draw with.
Be increasingly independent as they get dressed and undressed, for example, putting coats on and doing up zips.	Encourage children by helping them, but leaving them to do the last steps, such as pulling up their zip after you have started it off. Gradually reduce your help until the child can do each step on their own.

3 & 4-year-olds will be learning to:	Examples of how to support this:
Be increasingly independent in meeting their own care needs, e.g. brushing teeth, using the toilet, washing and drying their hands thoroughly. Make healthy choices about food, drink, activity and toothbrushing.	Talk to children about the importance of eating healthily and brushing their teeth. Consider how to support oral health. For example, some settings use a toothbrushing programme. Talk to children about why it's important to wash their hands carefully and throughout the day, including before they eat and after they've used the toilet.
Observation checkpoint	Most, but not all, children are reliably dry during the day by the age of 4. Support children who are struggling with toilet training, in partnership with their parents. Seek medical advice, if necessary, from a health visitor or GP.

Children in reception will be learning to:	Examples of how to support this:
Revise and refine the fundamental movement skills they have already acquired: - rolling - crawling - walking - jumping - running - hopping - skipping - climbing	Provide regular access to appropriate outdoor space. Ensure there is a range of surfaces to feel, move and balance on, such as grass, earth and bark chippings. Give children experience of carrying things up and down on different levels (slopes, hills and steps). Provide a choice of open-ended materials to play that allow for extended, repeated and regular practising of physical skills like lifting, carrying, pushing, pulling, constructing, stacking and climbing. Provide regular access to floor space indoors for movement. Ensure that spaces are accessible to children with varying confidence levels, skills and needs. Provide a wide range of activities to support a broad range of abilities. Allow less competent and confident children to spend time initially observing and listening, without feeling pressured to join in.

Children in reception will be learning to:	Examples of how to support this:
	Create low-pressure zones where less confident children can practise movement skills on their own, or with one or two others. Model precise vocabulary to describe movement and directionality, and encourage children to use it.
Progress towards a more fluent style of moving, with developing control and grace.	Provide children with regular opportunities to practise their movement skills alone and with others. Challenge children with further physical challenges when they are ready, such as climbing higher, running faster and jumping further. Encourage children to conclude movements in balance and stillness. Allow for time to be still and quiet. Suggestion: looking up at the sky, or sitting or lying in a den.
Develop the overall body strength, co-ordination, balance and agility needed to engage successfully with future physical education sessions and other physical disciplines including dance, gymnastics, sport and swimming.	Encourage children to be highly active and get out of breath several times every day. Provide opportunities for children to, spin, rock, tilt, fall, slide and bounce. Provide a range of wheeled resources for children to balance, sit or ride on, or pull and push. Two-wheeled balance bikes and pedal bikes without stabilisers, skateboards, wheelbarrows, prams and carts are all good options.
Develop their small motor skills so that they can use a range of tools competently, safely and confidently. Suggested tools: pencils for drawing and writing, paintbrushes, scissors, knives, forks and spoons.	Before teaching children the correct pencil grip and posture for writing, or how to use a knife and fork and cut with scissors, check: - that children have developed their upper arm and shoulder strength sufficiently: they don't need to move their shoulders as they move their hands and fingers - that they can move and rotate their lower arms and wrists independently Help children to develop the core strength and stability they need to support their small motor skills. Encourage and model tummy-crawling, crawling on all fours, climbing, pulling themselves up on a rope and hanging on monkey bars.

Children in reception will be learning to:	Examples of how to support this:
	Offer children activities to develop and further refine their small motor skills. Suggestions: threading and sewing, woodwork, pouring, stirring, dancing with scarves, using spray bottles, dressing and undressing dolls, planting and caring for plants, playing with small world toys, and making models with junk materials, construction kits and malleable materials like clay. Regularly review the equipment for children to develop their small motor skills. Is it appropriate for the different levels of skill and confidence of children in the class? Is it challenging for the most dexterous children? Continuously check how children are holding pencils for writing, scissors and knives and forks. Offer regular, gentle encouragement and feedback. With regular practice, the physical skills children need to eat with a knife and fork and develop an efficient handwriting style will become increasingly automatic.
Use their core muscle strength to achieve a good posture when sitting at a table or sitting on the floor.	Provide areas for sitting at a table that are quiet, purposeful and free of distraction. Give children regular, sensitive reminders about correct posture. Provide different chairs at the correct height for the range of children in the class, so that their feet are flat on the floor or a footrest. Provide different tables at the correct height for the range of children in the class. The table supports children's forearms. The top of the table is slightly higher than the height of the child's elbow flexed to 90 degrees.
Combine different movements with ease and fluency.	Create obstacle courses that demand a range of movements to complete, such as crawling through a tunnel, climbing onto a chair, jumping into a hoop and running and lying on a cushion. Provide opportunities to move that require quick changes of speed and direction. Suggestions: run around in a circle, stop, change direction and walk on your knees going the other way.

Children in reception will be learning to:	Examples of how to support this:
	Encourage precision and accuracy when beginning and ending movements.
Confidently and safely use a range of large and small apparatus indoors and outside, alone and in a group. Develop overall body-strength, balance, co-ordination and agility.	Encourage children to use a range of equipment. These might include: wheeled toys, wheelbarrows, tumbling mats, ropes to pull up on, spinning cones, tunnels, tyres, structures to jump on/off, den-making materials, logs and planks to balance on, A-frames and ladders, climbing walls, slides and monkey bars.
Further develop and refine a range of ball skills including: throwing, catching, kicking, passing, batting, and aiming. Develop confidence, competence, precision and accuracy when engaging in activities that involve a ball.	Provide a range of different sized 'balls' made out of familiar materials like socks, paper bags and jumpers that are softer and slower than real balls. Introduce full-sized balls when children are confident to engage with them. Introduce tennis balls, ping pong balls, beach balls and balloons. Introduce a range of resources used to bat, pat and hit a ball, modelling how to do this and giving children plenty of time for practice. Introduce children to balls games with teams, rules and targets when they have consolidated their ball skills.
Develop the foundations of a handwriting style which is fast, accurate and efficient.	Encourage children to draw freely. Engage children in structured activities: guide them in what to draw, write or copy. Teach and model correct letter formation. Continuously check the process of children's handwriting (pencil grip and letter formation, including directionality). Provide extra help and guidance when needed. Plan for regular repetition so that correct letter formation becomes automatic, efficient and fluent over time.

Children in reception will be learning to:	Examples of how to support this:
Know and talk about the different factors that support their overall health and wellbeing: - regular physical activity - healthy eating - toothbrushing - sensible amounts of 'screen time' - having a good sleep routine - being a safe pedestrian	Talk with children about exercise, healthy eating and the importance of sleep. Use picture books and other resources to explain the importance of the different aspects of a healthy lifestyle. Explain to children and model how to travel safely in their local environment, including: staying on the pavement, holding hands and crossing the road when walking, stopping quickly when scootering and cycling, and being sensitive to other pedestrians.
Further develop the skills they need to manage the school day successfully: - lining up and queuing - mealtimes - personal hygiene	Carefully explain some of the rules of lining up and queuing, such as not standing too close or touching others. Give children simple verbal and visual reminders. Celebrate, praise and reward children as they develop patience, turn-taking and self-control when they need to line up and wait. Teach and model for children how to eat with good manners in a group, taking turns and being considerate to others. Help individual children to develop good personal hygiene. Acknowledge and praise their efforts. Provide regular reminders about thorough handwashing and toileting. Work with parents and health visitors or the school nurse to help children who are not usually clean and dry through the day.

Literacy

It is crucial for children to develop a life-long love of reading. Reading consists of two dimensions: language comprehension and word reading. Language comprehension (necessary for both reading and writing) starts from birth. It only develops when adults talk with children about the world around them and the books (stories and non-fiction) they read with them, and enjoy rhymes, poems and songs together. Skilled word reading, taught later, involves both the speedy working out of the pronunciation of unfamiliar printed words (decoding) and the speedy recognition of familiar printed words. Writing involves transcription (spelling and handwriting) and composition (articulating ideas and structuring them in speech, before writing).

Statutory framework for the early years foundation stage: early adopter version

Birth to three - babies, toddlers and young children will be learning to:	**Examples of how to support this:**
Enjoy songs and rhymes, tuning in and paying attention. Join in with songs and rhymes, copying sounds, rhythms, tunes and tempo. Say some of the words in songs and rhymes. Copy finger movements and other gestures. Sing songs and say rhymes independently, for example, singing whilst playing.	Song and rhyme times can happen spontaneously throughout the day, indoors and outside, with individual children, in pairs or in small groups. You can make song and rhyme times engaging for young children by using a wide range of props or simple instruments. Children can choose the songs and rhymes they would like to join in with, using picture cards or by speaking You could learn songs and rhymes from parents. You could also teach parents the songs and rhymes you use in the setting, in order to support learning at home. Choose songs and rhymes which reflect the range of cultures and languages of children in the twenty-first century.
Enjoy sharing books with an adult. Pay attention and respond to the pictures or the words. Have favourite books and seek them out, to share with an	Provide enticing areas for sharing books, stocked with a wide range of high-quality books, matching the many different interests of children in the setting. Provide a comfortable place for sharing books, like a sofa. In warm weather, share books outside on a picnic rug or in small tents. Themed book areas can build on children's interests. Suggestions: relevant

Birth to three - babies, toddlers and young children will be learning to:	Examples of how to support this:
adult, with another child, or to look at alone. Repeat words and phrases from familiar stories. Ask questions about the book. Make comments and shares their own ideas. Develop play around favourite stories using props.	books close to small world play about dinosaurs, or cookbooks in the home corner. Help children to explore favourite books through linked activities. Suggestions: - visiting the park or the countryside to splash through puddles and squelch through mud for 'We're Going on a Bear Hunt' - going out to buy chillies for 'Lima's Red Hot Chilli' - dressing up clothes and small world play for favourite books
Notice some print, such as the first letter of their name, a bus or door number, or a familiar logo.	Point out print in the environment and talk about what it means. Suggestions: on a local walk, point out road signs, shop names and door numbers.
Enjoy drawing freely. Add some marks to their drawings, which they give meaning to. For example: "That says mummy." Make marks on their picture to stand for their name.	Provide a wide range of stimulating equipment to encourage children's mark-making. Suggestions: - large-scale sensory play, such as making marks with fingers in wet sand or in a tray of flour - using sticks and leaves to make marks during Forest school sessions - large brushes with paint or water - dragging streamers through puddles. - once large-muscle co-ordination is developing well, children can develop small-muscle co-ordination - playground chalk, smaller brushes, pencils and felt pens will support this.

3 & 4-year-olds will be learning to:	Examples of how to support this:
Understand the five key concepts about print: - print has meaning - print can have different purposes - we read English text from left to right and from top to bottom - the names of the different parts of a book	Draw children's attention to a wide range of examples of print with different functions. These could be a sign to indicate a bus stop or to show danger, a menu for choosing what you want to eat, or a logo that stands for a particular shop. When reading to children, sensitively draw their attention to the parts of the books, for example, the cover, the author, the page number. Show children how to handle books and to turn the pages one at a

3 & 4-year-olds will be learning to:	Examples of how to support this:
- page sequencing	time. Show children where the text is, and how English print is read left to right and top to bottom. Show children how sentences start with capital letters and end with full stops. Explain the idea of a 'word' to children, pointing out how some words are longer than others and how there is always a space before and after a word.
Develop their phonological awareness, so that they can: - spot and suggest rhymes - count or clap syllables in a word - recognise words with the same initial sound, such as money and mother	Help children tune into the different sounds in English by making changes to rhymes and songs, like: - changing a word so that there is still a rhyme: "Twinkle, twinkle yellow car" - making rhymes personal to children: "Hey diddle diddle, the cat and fiddle, the cow jumped over Haroon." Deliberately miss out a word in a rhyme, so the children have to fill it in: "Run, run, as fast as you **can,** you can't catch me I'm the gingerbread —." Use magnet letters to spell a word ending like 'at'. Encourage children to put other letters in front to create rhyming words like 'hat' and 'cat'.
Engage in extended conversations about stories, learning new vocabulary.	Choose books which reflect diversity. Regular sharing of books and discussion of children's ideas and responses (dialogic reading) helps children to develop their early enjoyment and understanding of books. Simple picture books, including those with no text, can be powerful ways of learning new vocabulary (for example, naming what's in the picture). More complex stories will help children to learn a wider range of vocabulary. This type of vocabulary is not in everyday use, but occurs frequently in books and other contexts. Examples include: 'caterpillar', 'enormous', 'forest', 'roar' and 'invitation'.
Use some of their print and letter knowledge in their early writing. For example: writing a pretend shopping list that starts at the top of the page; writing 'm' for mummy. Write some or all of their name.	Motivate children to write by providing opportunities in a wide range of ways. Suggestions: clipboards outdoors, chalks for paving stones, boards and notepads in the home corner. Children enjoy having a range of pencils, crayons, chalks and pens to choose from. Apps on tablets enable children to mix marks, photos and video to express meanings and tell their own stories. Children are also motivated by simple home-made books, different coloured paper and paper decorated with fancy frames.

3 & 4-year-olds will be learning to:	Examples of how to support this:
Write some letters accurately.	Help children to learn to form their letters accurately. First, they need a wide-ranging programme of physical skills development, inside and outdoors. Include large-muscle co-ordination: whole body, leg, arm and foot. This can be through climbing, swinging, messy play and parachute games etc. Plan for small-muscle co-ordination: hands and fingers. This can be through using scissors, learning to sew, eating with cutlery, using small brushes for painting and pencils for drawing. Children also need to know the language of direction ('up', 'down', 'round', 'back' etc).

Children in reception will be learning to:	Examples of how to support this:
Read individual letters by saying the sounds for them.	Help children to read the sounds speedily. This will make sound-blending easier.
Blend sounds into words, so that they can read short words made up of known letter–sound correspondences.	Ask children to work out the word you say in sounds: for example, h-a-t > hat; sh-o-p > shop. Show how to say sounds for the letters from left to right and blend them, for example, big, stamp.
Read some letter groups that each represent one sound and say sounds for them.	Help children to become familiar with letter groups, such as 'th', 'sh', 'ch', 'ee' 'or' 'igh'. Provide opportunities for children to read words containing familiar letter groups: 'that', 'shop', 'chin', 'feet', 'storm', 'night'. Listen to children read some longer words made up of letter-sound correspondences they know: 'rabbit', 'himself', 'jumping'.
Read a few common exception words matched to the school's phonic programme.	Note correspondences between letters and sounds that are unusual or that they have not yet been taught, such as 'do', 'said', 'were'.
Read simple phrases and sentences made up of words with known letter–sound correspondences and, where necessary, a few exception words.	Listen to children read aloud, ensuring books are consistent with their developing phonic knowledge. Do not include words that include letter-sound correspondences that children cannot yet read, or exception words that have not been taught. Children should not be required to use other strategies to work out words.

Children in reception will be learning to:	**Examples of how to support this:**
Re-read these books to build up their confidence in word reading, their fluency and their understanding and enjoyment.	Make the books available for children to share at school and at home. Avoid asking children to read books at home they cannot yet read.
Form lower-case and capital letters correctly.	Teach formation as they learn the sounds for each letter using a memorable phrase.
Spell words by identifying the sounds and then writing the sound with letter/s.	Show children how to touch each finger as they say each sound. For exception words such as 'the' and 'said', help children identify the sound that is tricky to spell.
Write short sentences with words with known sound-letter correspondences using a capital letter and full stop.	Support children to form the complete sentence before writing. Help children memorise the sentence before writing by saying it aloud. Only ask children to write sentences when they have sufficient knowledge of letter-sound correspondences.
Re-read what they have written to check that it makes sense.	Model how you read and re-read your own writing to check it makes sense.

Mathematics

Developing a strong grounding in number is essential so that all children develop the necessary building blocks to excel mathematically. Children should be able to count confidently, develop a deep understanding of the numbers to 10, the relationships between them and the patterns within those numbers. By providing frequent and varied opportunities to build and apply this understanding - such as using manipulatives, including small pebbles and tens frames for organising counting - children will develop a secure base of knowledge and vocabulary from which mastery of mathematics is built. In addition, it is important that the curriculum includes rich opportunities for children to develop their spatial reasoning skills across all areas of mathematics including shape, space and measures. It is important that children develop positive attitudes and interests in mathematics, look for patterns and relationships, spot connections, 'have a go', talk to adults and peers about what they notice and not be afraid to make mistakes.

Statutory framework for the early years foundation stage: early adopter version

Birth to three - babies, toddlers and young children will be learning to:	Examples of how to support this:
Combine objects like stacking blocks and cups. Put objects inside others and take them out again.	Encourage babies and young toddlers to play freely with a wide range of objects - toddlers engage spontaneously in mathematics during nearly half of every minute of free play. Suggestions: when appropriate, sensitively join in and comment on: - interestingly shaped objects like vegetables, wooden pegs, spoons, pans, corks, cones, balls - pots and pans, boxes and objects to put in them, shape sorters - stacking cups: hiding one, building them into a tower, nesting them and lining them up
Take part in finger rhymes with numbers. React to changes of amount in a group of up to three items.	Use available opportunities, including feeding and changing times for finger-play, outdoors and inside, such as 'Round and round the garden'. Sing finger rhymes which involve hiding and returning, like 'Two little dicky birds'.
Compare amounts, saying 'lots', 'more' or 'same'. Develop counting-like behaviour, such as making sounds, pointing or saying some numbers in sequence.	Draw attention to changes in amounts, for example, by adding more bricks to a tower, or eating things up Offer repeated experiences with the counting sequence in meaningful and varied contexts, outside and indoors. Suggestions: count fingers and toes, stairs, toys, food items, sounds and actions.

Birth to three - babies, toddlers and young children will be learning to:	Examples of how to support this:
Count in everyday contexts, sometimes skipping numbers - '1-2-3-5.'	Help children to match their counting words with objects. Suggestions: move a piece of apple to one side once they have counted it. If children are saying one number word for each object, it isn't always necessary to correct them if they skip a number. Learning to count accurately takes a long time and repeated experience. Confidence is important.
Climb and squeeze themselves into different types of spaces. Build with a range of resources. Complete inset puzzles.	Describe children's climbing, tunnelling and hiding using spatial words like 'on top of', 'up', 'down' and 'through'. Provide blocks and boxes to play freely with and build with, indoors and outside. Provide inset puzzles and jigsaws at different levels of difficulty.
Compare sizes, weights etc. using gesture and language - 'bigger/little/smaller', 'high/low', 'tall', 'heavy'.	Use the language of size and weight in everyday contexts. Provide objects with marked differences in size to play freely with. Suggestions: dolls' and adult chairs, tiny and big bears, shoes, cups and bowls, blocks and containers.
Notice patterns and arrange things in patterns.	Provide patterned material – gingham, polka dots, stripes etc. – and small objects to arrange in patterns. Use words like 'repeated' and 'the same' over and over.

3 & 4-year-olds will be learning to:	Examples of how to support this:
Develop fast recognition of up to 3 objects, without having to count them individually ('subitising'). Recite numbers past 5. Say one number for each item in order: 1,2,3,4,5.	Point to small groups of two or three objects: "Look, there are two!" Occasionally ask children how many there are in a small set of two or three. Regularly say the counting sequence, in a variety of playful contexts, inside and outdoors, forwards and backwards, sometimes going to high numbers. For example: hide and seek, rocket-launch countdowns. Count things and then repeat the last number. For example: "1, 2, 3 – **3 cars**". Point out the number of things whenever possible; so, rather than just

3 & 4-year-olds will be learning to:	Examples of how to support this:
Know that the last number reached when counting a small set of objects tells you how many there are in total ('cardinal principle'). Show 'finger numbers' up to 5. Link numerals and amounts: for example, showing the right number of objects to match the numeral, up to 5.	'chairs', 'apples' or 'children', say 'two chairs', 'three apples', 'four children'. Ask children to get you a number of things, and emphasise the total number in your conversation with the child. Use small numbers to manage the learning environment. Suggestions: have a pot labelled '5 pencils' or a crate for '3 trucks'. Draw children's attention to these throughout the session and especially at tidy-up time: "How many pencils should be in this pot?" or "How many have we got?" etc.
Experiment with their own symbols and marks as well as numerals. Solve real world mathematical problems with numbers up to 5. Compare quantities using language: 'more than', 'fewer than'.	Encourage children in their own ways of recording (for example) how many balls they managed to throw through the hoop. Provide numerals nearby for reference. Suggestions: wooden numerals in a basket or a number track on the fence. Discuss mathematical ideas throughout the day, inside and outdoors. Suggestions: - "I think Adam has got more crackers…" - support children to solve problems using fingers, objects and marks: "There are four of you, but there aren't enough chairs…." - draw children's attention to differences and changes in amounts, such as those in stories like 'The Enormous Turnip'.
Talk about and explore 2D and 3D shapes (for example, circles, rectangles, triangles and cuboids) using informal and mathematical language: 'sides', 'corners'; 'straight', 'flat', 'round'.	Encourage children to play freely with blocks, shapes, shape puzzles and shape-sorters. Sensitively support and discuss questions like: "What is the same and what is different?" Encourage children to talk informally about shape properties using words like 'sharp corner', 'pointy' or 'curvy'. Talk about shapes as you play with them: "We need a piece with a straight edge."
Understand position through words alone – for example,	Discuss position in real contexts. Suggestions: how to shift the leaves **off** a path, or sweep water away **down** the drain.

3 & 4-year-olds will be learning to:	Examples of how to support this:
"The bag is under the table," – with no pointing. Describe a familiar route. Discuss routes and locations, using words like 'in front of' and 'behind'.	Use spatial words in play, including 'in', 'on', 'under', 'up', 'down', 'besides' and 'between'. Suggestion: "Let's put the troll under the bridge and the billy goat beside the stream." Take children out to shops or the park: recall the route and the order of things seen on the way. Set up obstacle courses, interesting pathways and hiding places for children to play with freely. When appropriate, ask children to describe their route and give directions to each other. Provide complex train tracks, with loops and bridges, or water-flowing challenges with guttering that direct the flow to a water tray, for children to play freely with. Read stories about journeys, such as 'Rosie's Walk'.
Make comparisons between objects relating to size, length, weight and capacity.	Provide experiences of size changes. Suggestions: "Can you make a puddle larger?", "When you squeeze a sponge, does it stay small?", "What happens when you stretch dough, or elastic?" Talk with children about their everyday ways of comparing size, length, weight and capacity. Model more specific techniques, such as lining up ends of lengths and straightening ribbons, discussing accuracy: "Is it **exactly**…?"
Select shapes appropriately: flat surfaces for building, a triangular prism for a roof etc. Combine shapes to make new ones - an arch, a bigger triangle etc.	Provide a variety of construction materials like blocks and interlocking bricks. Provide den-making materials. Allow children to play freely with these materials, outdoors and inside. When appropriate, talk about the shapes and how their properties suit the purpose. Provide shapes that combine to make other shapes, such as pattern blocks and interlocking shapes, for children to play freely with. When appropriate, discuss the different designs that children make.

3 & 4-year-olds will be learning to:	Examples of how to support this:
	Occasionally suggest challenges, so that children build increasingly more complex constructions. Use tidy-up time to match blocks to silhouettes or fit things in containers, describing and naming shapes. Suggestion: "Where does this triangular one /cylinder /cuboid go?"
Talk about and identify the patterns around them. For example: stripes on clothes, designs on rugs and wallpaper. Use informal language like 'pointy', 'spotty', 'blobs' etc. Extend and create ABAB patterns – stick, leaf, stick, leaf. Notice and correct an error in a repeating pattern. Begin to describe a sequence of events, real or fictional, using words such as 'first', 'then...'	Provide patterns from different cultures, such as fabrics. Provide a range of natural and everyday objects and materials, as well as blocks and shapes, for children to play with freely and to make patterns with. When appropriate, encourage children to continue patterns and spot mistakes. Engage children in following and inventing movement and music patterns, such as clap, clap, stamp. Talk about patterns of events, in cooking or getting dressed. Suggestions: - 'First', 'then', 'after', 'before' - "Every day we..." - "Every evening we..." Talk about the sequence of events in stories. Use vocabulary like 'morning', 'afternoon', 'evening' and 'night-time', 'earlier', 'later', 'too late', 'too soon', 'in a minute'. Count down to forthcoming events on the calendar in terms of number of days or sleeps. Refer to the days of the week, and the day before or day after, 'yesterday' and 'tomorrow'.

Children in reception will be learning to:	Examples of how to support this:
Count objects, actions and sounds.	Develop the key skills of counting objects including saying the numbers in order and matching one number name to each item. Say how many there are after counting - for example, "...6, 7, 8. There are **8 balls**" - to help children appreciate that the last number of the count indicates the total number of the group. This is the cardinal counting principle. Say how many there might be before you count to give a purpose to counting: "I think there are about 8. Shall we count to see?" Count out a smaller number from a larger group: "Give me seven..." Knowing when to stop shows that children understand the cardinal principle. Build counting into everyday routines such as register time, tidying up, lining up or counting out pieces of fruit at snack time. Sing counting songs and number rhymes, and read stories that involve counting. Play games which involve counting. Identify children who have had less prior experience of counting, and provide additional opportunities for counting practice.
Subitise.	Show small quantities in familiar patterns (for example, dice) and random arrangements. Play games which involve quickly revealing and hiding numbers of objects. Put objects into five frames and then ten frames to begin to familiarise children with the tens structure of the number system. Prompt children to subitise first when enumerating groups of up to 4 or 5 objects: "I don't think we need to count those. They are in a square shape so there must be 4." Count to check.

Children in reception will be learning to:	Examples of how to support this:
	Encourage children to show a number of fingers 'all at once', without counting.
Link the number symbol (numeral) with its cardinal number value.	Display numerals in order alongside dot quantities or tens frame arrangements. Play card games such as snap or matching pairs with cards where some have numerals and some have dot arrangements. Discuss the different ways children might record quantities (for example, scores in games), such as tallies, dots and using numeral cards.
Count beyond ten.	Count verbally beyond 20, pausing at each multiple of 10 to draw out the structure, for instance when playing hide and seek, or to time children getting ready. Provide images such as number tracks, calendars and hundred squares indoors and out, including painted on the ground, so children become familiar with two-digit numbers and can start to spot patterns within them.
Compare numbers.	Provide collections to compare, starting with a very different number of things. Include more small things and fewer large things, spread them out and bunch them up, to draw attention to the number not the size of things or the space they take up. Include groups where the number of items is the same. Use vocabulary: 'more than', 'less than', 'fewer', 'the same as', 'equal to'. Encourage children to use these words as well. Distribute items evenly, for example: "Put 3 in each bag," or give the same number of pieces of fruit to each child. Make deliberate mistakes to provoke discussion. Tell a story about a character distributing snacks unfairly and invite children to make sure everyone has the same.
Understand the 'one more than/one less than' relationship between consecutive numbers.	Make predictions about what the outcome will be in stories, rhymes and songs if one is added, or if one is taken away.

Children in reception will be learning to:	**Examples of how to support this:**
	Provide 'staircase' patterns which show that the next counting number includes the previous number plus one.
Explore the composition of numbers to 10.	Focus on composition of 2, 3, 4 and 5 before moving onto larger numbers Provide a range of visual models of numbers: for example, six as double three on dice, or the fingers on one hand and one more, or as four and two with ten frame images. Model conceptual subitising: "Well, there are three here and three here, so there must be six." Emphasise the parts within the whole: "There were 8 eggs in the incubator. Two have hatched and 6 haven't yet hatched." Plan games which involve partitioning and recombining sets. For example, throw 5 beanbags, aiming for a hoop. How many go in and how many don't?
Automatically recall number bonds for numbers 0–10.	Have a sustained focus on each number to 10. Make visual and practical displays in the classroom showing the different ways of making numbers to 10 so that children can refer to these. Play hiding games with a number of objects in a box, under a cloth, in a tent, in a cave, etc.: "Seven went in the tent and 2 came out. I wonder how many are still in there?" Intentionally give children the wrong number of things. For example: ask each child to plant 4 seeds then give them 1, 2 or 3. "I've only got 1 seed, I need 3 more." Spot and use opportunities for children to apply number bonds: "There are 6 of us but only 2 clipboards. How many more do we need?" Place objects into a five frame and talk about how many spaces are filled and unfilled.

Children in reception will be learning to:	**Examples of how to support this:**
Select, rotate and manipulate shapes in order to develop spatial reasoning skills.	Provide high-quality pattern and building sets, including pattern blocks, tangrams, building blocks and magnetic construction tiles, as well as found materials. Challenge children to copy increasingly complex 2D pictures and patterns with these 3D resources, guided by knowledge of learning trajectories: "I bet you can't add an arch to that," or "Maybe tomorrow someone will build a staircase." Teach children to solve a range of jigsaws of increasing challenge.
Compose and decompose shapes so that children recognise a shape can have other shapes *within* it, just as numbers can.	Investigate how shapes can be combined to make new shapes: for example, two triangles can be put together to make a square. Encourage children to predict what shapes they will make when paper is folded. Wonder aloud how many different ways there are to make a hexagon with pattern blocks. Find 2D shapes within 3D shapes, including through printing or shadow play.
Continue, copy and create repeating patterns.	Make patterns with varying rules (including AB, ABB and ABBC) and objects and invite children to continue the pattern. Make a deliberate mistake and discuss how to fix it.
Compare length, weight and capacity.	Model comparative language using 'than' and encourage children to use this vocabulary. For example: "This is heavier than that." Ask children to make and test predictions. "What if we pour the jugful into the teapot? Which holds more?"

Understanding the world

Understanding the world involves guiding children to make sense of their physical world and their community. The frequency and range of children's personal experiences increases their knowledge and sense of the world around them – from visiting parks, libraries and museums to meeting important members of society such as police officers, nurses and firefighters. In addition, listening to a broad selection of stories, non-fiction, rhymes and poems will foster their understanding of our culturally, socially, technologically and ecologically diverse world. As well as building important knowledge, this extends their familiarity with words that support understanding across domains. Enriching and widening children's vocabulary will support later reading comprehension.

Statutory framework for the early years foundation stage: early adopter version

Birth to three - babies, toddlers and young children will be learning to:	Examples of how to support this:
Repeat actions that have an effect. Explore materials with different properties. Explore natural materials, indoors and outside.	Encourage babies' explorations and movements, such as touching their fingers and toes. Show delight at their kicking and waving. Provide open-ended play materials inside and outdoors. Suggestion: Treasure Baskets for repeated exploration of textures, sounds, smells and tastes. Offer lots of different textures for exploration with fingers, feet and whole body. Suggestions: wet and dry sand, water, paint and playdough.
Explore and respond to different natural phenomena in their setting and on trips.	Encourage toddlers and young children to enjoy and explore the natural world. Suggestions: - standing in the rain with wellies and umbrellas - walking through tall grass - splashing in puddles - seeing the spring daffodils and cherry blossom - looking for worms and minibeasts - visiting the beach and exploring the sand, pebbles and paddling in the sea Encourage children's exploration, curiosity, appreciation and respect for living things. Suggestions:

Birth to three - babies, toddlers and young children will be learning to:	Examples of how to support this:
	- sharing the fascination of a child who finds woodlice teeming under an old log - modelling the careful handling of a worm and helping children return it to the dug-up soil - carefully planting, watering and looking after plants they have grown from seeds Encourage children to bring natural materials into the setting, such as leaves and conkers picked up from the pavement or park during autumn.
Make connections between the features of their family and other families.	Be open to children talking about differences and what they notice. For example, when children ask questions like: "Why do you wear a scarf around your head?" or "How come your hair feels different to mine?" Point out the similarities between different families, as well as discussing differences.
Notice differences between people.	Model positive attitudes about the differences between people including differences in race and religion. Support children's acceptance of difference. Have resources which include: - positive images of people who are disabled - books and play materials that reflect the diversity of life in modern Britain including racial and religious diversity. - materials which confront gender stereotypes

3 & 4-year-olds will be learning to:	Examples of how to support this:
Use all their senses in hands-on exploration of natural materials. Explore collections of materials with similar and/or different properties. Talk about what they see, using a wide vocabulary.	Provide interesting natural environments for children to explore freely outdoors. Make collections of natural materials to investigate and talk about. Suggestions: - contrasting pieces of bark - different types of leaves and seeds - different types of rocks - different shells and pebbles from the beach Provide equipment to support these investigations. Suggestions: magnifying glasses or a tablet with a magnifying app.

3 & 4-year-olds will be learning to:	Examples of how to support this:
	Encourage children to talk about what they see. Model observational and investigational skills. Ask out loud: "I wonder if…?" Plan and introduce new vocabulary, encouraging children to use it to discuss their findings and ideas.
Begin to make sense of their own life-story and family's history.	Spend time with children talking about photos and memories. Encourage children to retell what their parents told them about their life story and family.
Show interest in different occupations.	Invite different people to visit from a range of occupations, such as a plumber, a farmer, a vet, a member of the emergency services or an author. Plan and introduce new vocabulary related to the occupation, and encourage children to use it in their speech and play. Consider opportunities to challenge gender and other stereotypes.
Explore how things work.	Provide mechanical equipment for children to play with and investigate. Suggestions: wind-up toys, pulleys, sets of cogs with pegs and boards.
Plant seeds and care for growing plants. Understand the key features of the life cycle of a plant and an animal. Begin to understand the need to respect and care for the natural environment and all living things.	Show and explain the concepts of growth, change and decay with natural materials. Suggestions: - plant seeds and bulbs so children observe growth and decay over time - observe an apple core going brown and mouldy over time - help children to care for animals and take part in first-hand scientific explorations of animal life cycles, such as caterpillars or chick eggs. Plan and introduce new vocabulary related to the exploration. Encourage children to use it in their discussions, as they care for living things.

3 & 4-year-olds will be learning to:	Examples of how to support this:
	Encourage children to refer to books, wall displays and online resources. This will support their investigations and extend their knowledge and ways of thinking.
Explore and talk about different forces they can feel.	Draw children's attention to forces. Suggestions: - how the water pushes up when they try to push a plastic boat under it - how they can stretch elastic, snap a twig, but can't bend a metal rod - magnetic attraction and repulsion Plan and introduce new vocabulary related to the exploration, and encourage children to use it.
Talk about the differences between materials and changes they notice.	Provide children with opportunities to change materials from one state to another. Suggestions: - cooking – combining different ingredients, and then cooling or heating (cooking) them - melting – leave ice cubes out in the sun, see what happens when you shake salt onto them (children should not touch to avoid danger of frostbite) Explore how different materials sink and float. Explore how you can shine light through some materials, but not others. Investigate shadows. Plan and introduce new vocabulary related to the exploration, and encourage children to use it.
Continue developing positive attitudes about the differences between people.	Ensure that resources reflect the diversity of life in modern Britain. Encourage children to talk about the differences they notice between people, whilst also drawing their attention to similarities between different families and communities.

3 & 4-year-olds will be learning to:	Examples of how to support this:
	Answer their questions and encourage discussion. Suggestion: talk positively about different appearances, skin colours and hair types. Celebrate and value cultural, religious and community events and experiences. Help children to learn each other's names, modelling correct pronunciation.
Know that there are different countries in the world and talk about the differences they have experienced or seen in photos.	Practitioners can create books and displays about children's families around the world, or holidays they have been on. Encourage children to talk about each other's families and ask questions. Use a diverse range of props, puppets, dolls and books to encourage children to notice and talk about similarities and differences.

Children in reception will be learning to:	Examples of how to support this:
Talk about members of their immediate family and community.	During dedicated talk time, listen to what children say about their family. Share information about your own family, giving children time to ask questions or make comments. Encourage children to share pictures of their family and listen to what they say about the pictures. Using examples from real life and from books, show children how there are many different families.
Name and describe people who are familiar to them.	Talk about people that the children may have come across within their community, such as the police, the fire service, doctors and teachers. Listen to what children say about their own experiences with people who are familiar to them.

Children in reception will be learning to:	Examples of how to support this:
Comment on images of familiar situations in the past.	Present children with pictures, stories, artefacts and accounts from the past, explaining similarities and differences. Offer hands-on experiences that deepen children's understanding, such as visiting a local area that has historical importance. Show images of familiar situations in the past, such as homes, schools, and transport. Look for opportunities to observe children talking about experiences that are familiar to them and how these may have differed in the past. Offer opportunities for children to begin to organise events using basic chronology, recognising that things happened before they were born.
Compare and contrast characters from stories, including figures from the past.	Frequently share texts, images, and tell oral stories that help children begin to develop an understanding of the past and present. Feature fictional and non-fictional characters from a range of cultures and times in storytelling, listen to what children say about them. Draw out common themes from stories, such as bravery, difficult choices and kindness, and talk about children's experiences with these themes. In addition to storytelling, introduce characters, including those from the past using songs, poems, puppets, role play and other storytelling methods.
Draw information from a simple map.	Draw children's attention to the immediate environment, introducing and modelling new vocabulary where appropriate. Familiarise children with the name of the road, and or village/town/city the school is located in. Look at aerial views of the school setting, encouraging children to comment on what they

Children in reception will be learning to:	Examples of how to support this:
	notice, recognising buildings, open space, roads and other simple features. Offer opportunities for children to choose to draw simple maps of their immediate environment, or maps from imaginary story settings they are familiar with.
Understand that some places are special to members of their community.	Name and explain the purpose of places of worship and places of local importance to the community to children, drawing on their own experiences where possible Take children to places of worship and places of local importance to the community. Invite visitors from different religious and cultural communities into the classroom to share their experiences with children.
Recognise that people have different beliefs and celebrate special times in different ways.	Weave opportunities for children to engage with religious and cultural communities and their practices throughout the curriculum at appropriate times of the year. Help children to begin to build a rich bank of vocabulary with which to describe their own lives and the lives of others.
Recognise some similarities and differences between life in this country and life in other countries.	Teach children about places in the world that contrast with locations they know well. Use relevant, specific vocabulary to describe contrasting locations. Use images, video clips, shared texts and other resources to bring the wider world into the classroom. Listen to what children say about what they see. Avoid stereotyping and explain how children's lives in other countries may be similar or different in terms of

Children in reception will be learning to:	Examples of how to support this:
	how they travel to school, what they eat, where they live, and so on.
Explore the natural world around them.	Provide children with have frequent opportunities for outdoor play and exploration. Encourage interactions with the outdoors to foster curiosity and give children freedom to touch, smell and hear the natural world around them during hands-on experiences. Create opportunities to discuss how we care for the natural world around us. Offer opportunities to sing songs and join in with rhymes and poems about the natural world. After close observation, draw pictures of the natural world, including animals and plants. Observe and interact with natural processes, such as ice melting, a sound causing a vibration, light travelling through transparent material, an object casting a shadow, a magnet attracting an object and a boat floating on water.
Describe what they see, hear and feel whilst outside.	Encourage focused observation of the natural world. Listen to children describing and commenting on things they have seen whilst outside, including plants and animals. Encourage positive interaction with the outside world, offering children a chance to take supported risks, appropriate to themselves and the environment within which they are in. Name and describe some plants and animals children are likely to see, encouraging children to recognise familiar plants and animals whilst outside.

Children in reception will be learning to:	**Examples of how to support this:**
Recognise some environments that are different to the one in which they live.	Teach children about a range of contrasting environments within both their local or national region. Model the vocabulary needed to name specific features of the natural world, both natural and man-made. Share non-fiction texts that offer an insight into contrasting environments. Listen to how children communicate their understanding of their own environment and contrasting environments through conversation and in play.
Understand the effect of changing seasons on the natural world around them.	Guide children's understanding by draw children's attention to the weather and seasonal features. Provide opportunities for children to note and record the weather. Select texts to share with the children about the changing seasons. Throughout the year, take children outside to observe the natural world and encourage children to observe how animals behave differently as the seasons change. Look for children incorporating their understanding of the seasons and weather in their play.

Expressive arts and design

The development of children's artistic and cultural awareness supports their imagination and creativity. It is important that children have regular opportunities to engage with the arts, enabling them to explore and play with a wide range of media and materials. The quality and variety of what children see, hear and participate in is crucial for developing their understanding, self-expression, vocabulary and ability to communicate through the arts. The frequency, repetition and depth of their experiences are fundamental to their progress in interpreting and appreciating what they hear, respond to and observe.

Statutory framework for the early years foundation stage: early adopter version

Birth to three - babies, toddlers and young children will be learning to:	Examples of how to support this:
Show attention to sounds and music. Respond emotionally and physically to music when it changes. Move and dance to music. Anticipate phrases and actions in rhymes and songs, like 'Peepo'. Explore their voices and enjoy making sounds.	Babies are born ready to enjoy and make music from birth. Stimulate their enjoyment of music through singing and playing musical and singing games which are attuned to the baby. Provide babies, toddlers and young children with a range of different types of singing, sounds and music from diverse cultures. Music and singing can be live as well as pre-recorded. Play and perform music with different: - dynamics (loud/quiet) - tempo (fast/slow) - pitch (high/low) - rhythms (pattern of sound)
Join in with songs and rhymes, making some sounds. Make rhythmical and repetitive sounds. Explore a range of sound-makers and instruments and play them in different ways.	Introduce children to songs, including songs to go with routines. Suggestion: when washing hands, sing "This is the ways we wash our hands…". Provide children with instruments and with 'found objects'. Suggestions: tapping a bottle onto the table or running a twig along a fence. Encourage children to experiment with different ways of playing instruments.
Notice patterns with strong contrasts and be attracted by patterns resembling the human face.	Ensure that the physical environment includes objects and materials with different patterns, colours, tones and textures for babies and young children to explore.

Birth to three - babies, toddlers and young children will be learning to:	Examples of how to support this:
Start to make marks intentionally. Explore paint, using fingers and other parts of their bodies as well as brushes and other tools. Express ideas and feelings through making marks, and sometimes give a meaning to the marks they make.	Stimulate babies' and toddlers' early interest in making marks. Offer a wide range of different materials and encourage children to make marks in different ways. Suggestions: - invite them to submerge their fingers in cornflour - play with a stick in the mud - place hands and feet in paint - use tablets or computers - introduce colour names
Enjoy and take part in action songs, such as 'Twinkle, Twinkle Little Star'.	Introduce children to a broad selection of action songs from different cultures and languages. Sing songs regularly so that children learn the words, melody and actions off by heart. Encourage children to accompany action songs. They can do this with their own movements or by playing instruments.
Start to develop pretend play, pretending that one object represents another. For example, a child holds a wooden block to her ear and pretends it's a phone.	Children generally start to understand the difference between pretend and real from around the age of 2. Help children to develop their pretend play by modelling, sensitively joining in and helping them to elaborate it. Suggestion: help to develop a child's home corner play of feeding a 'baby', by suggesting a nappy-change and then a song as you settle the 'baby' to sleep.
Explore different materials, using all their senses to investigate them. Manipulate and play with different materials. Use their imagination as they consider what they can do with different materials. Make simple models which express their ideas.	Stimulate young children's interest in modelling. Suggestions: provide a wide range of found materials ('junk') as well as blocks, clay, soft wood, card, off-cuts of fabrics and materials with different textures. Provide appropriate tools and joining methods for the materials offered. Encourage young children to explore materials/ resources finding out what they are/what they can do, and decide how they want to use them.

3 & 4-year-olds will be learning to:	Examples of how to support this:
Take part in simple pretend play, using an object to represent something else even though they are not similar. Begin to develop complex stories using small world equipment like animal sets, dolls and dolls houses etc. Make imaginative and complex 'small worlds' with blocks and construction kits, such as a city with different buildings and a park.	Children generally start to develop pretend play with 'rules' when they are 3 or 4 years old. Suggestion: offer pinecones in the home corner for children to pour into pans and stir like pasta. Some rules are self-created (the pole is now a horse, or the pinecones are now pasta in the pot). Other rules are group-created (to play in the home corner, you must accept the rule that one of your friends is pretending to be a baby). Provide lots of flexible and open-ended resources for children's imaginative play. Help children to negotiate roles in play and sort out conflicts. Notice children who are not taking part in pretend play, and help them to join in.
Explore different materials freely, in order to develop their ideas about how to use them and what to make. Develop their own ideas and then decide which materials to use to express them. Join different materials and explore different textures.	Offer opportunities to explore scale. Suggestions: - long strips of wallpaper - child size boxes - different surfaces to work on e.g. paving, floor, tabletop or easel Listen and understand what children want to create before offering suggestions. Invite artists, musicians and craftspeople into the setting, to widen the range of ideas which children can draw on. Suggestions: glue and masking tape for sticking pieces of scrap materials onto old cardboard boxes, hammers and nails, glue guns, paperclips and fasteners.
Create closed shapes with continuous lines, and begin to use these shapes to represent objects.	Help children to develop their drawing and model-making. Encourage them to develop their own creative ideas. Spend sustained time alongside them. Show interest in the meanings children give to

3 & 4-year-olds will be learning to:	Examples of how to support this:
Draw with increasing complexity and detail, such as representing a face with a circle and including details. Use drawing to represent ideas like movement or loud noises. Show different emotions in their drawings and paintings, like happiness, sadness, fear etc. Explore colour and colour-mixing. Show different emotions in their drawings – happiness, sadness, fear etc.	their drawings and models. Talk together about these meanings. Encourage children to draw from their imagination and observation. Help children to add details to their drawings by selecting interesting objects to draw, and by pointing out key features to children and discussing them. Talk to children about the differences between colours. Help them to explore and refine their colour-mixing - for example: "How does blue become green?" Introduce children to the work of artists from across times and cultures. Help them to notice where features of artists' work overlap with the children's, for example in details, colour, movement or line.
Listen with increased attention to sounds. Respond to what they have heard, expressing their thoughts and feelings.	Help children to develop their listening skills through a range of active listening activities. Notice 'how' children listen well, for example: listening whilst painting or drawing, or whilst moving. Play, share and perform a wide variety of music and songs from different cultures and historical periods. Play sound-matching games.
Remember and sing entire songs. Sing the pitch of a tone sung by another person ('pitch match'). Sing the melodic shape (moving melody, such as up	When teaching songs to children be aware of your own pitch (high/low). Children's voices are higher than adult voices. When supporting children to develop their singing voice use a limited pitch range. For example, 'Rain rain' uses a smaller pitch (high/low) range than many traditional nursery rhymes. Children's singing voices and their ability to control them is developing. Encourage them to use their 'singing' voice: when asked to sing loudly, children often shout.

3 & 4-year-olds will be learning to:	Examples of how to support this:
and down, down and up) of familiar songs. Create their own songs, or improvise a song around one they know.	Sing slowly, so that children clearly hear the words and the melody of the song. Use songs with and without words – children may pitch-match more easily without words. Try using one-syllable sounds such as 'ba'. Clap or tap to the pulse of songs or music, and encourage children to do this.
Play instruments with increasing control to express their feelings and ideas.	Offer children a wide range of different instruments, from a range of cultures. This might also include electronic keyboards and musical apps on tablets. Encourage children to experiment with different ways of playing instruments. Listen carefully to their music making and value it. Suggestion: record children's pieces, play the pieces back to the children and include them in your repertoire of music played in the setting.

Children in reception will be learning to:	Examples of how to support this:
Explore, use and refine a variety of artistic effects to express their ideas and feelings. Return to and build on their previous learning, refining ideas and developing their ability to represent them. Create collaboratively, sharing ideas, resources and skills.	Teach children to develop their colour-mixing techniques to enable them to match the colours they see and want to represent, with step-by-step guidance when appropriate. Provide opportunities to work together to develop and realise creative ideas. Provide children with a range of materials for children to construct with. Encourage them to think about and discuss what they want to make. Discuss problems and how they might be solved as they arise. Reflect with children on how they have achieved their aims. Teach children different techniques for joining materials, such as how to use adhesive tape and different sorts of glue. Provide a range of materials and tools and teach children to use them with care and precision. Promote

Children in reception will be learning to:	Examples of how to support this:
	independence, taking care not to introduce too many new things at once. Encourage children to notice features in the natural world. Help them to define colours, shapes, texture and smells in their own words. Discuss children's responses to what they see. Visit galleries and museums to generate inspiration and conversation about art and artists.
Listen attentively, move to and talk about music, expressing their feelings and responses.	Give children an insight into new musical worlds. Introduce them to different kinds of music from across the globe, including traditional and folk music from Britain. Invite musicians in to play music to children and talk about it. Encourage children to listen attentively to music. Discuss changes and patterns as a piece of music develops.
Watch and talk about dance and performance art, expressing their feelings and responses.	Offer opportunities for children to go to a live performance, such as a pantomime, play, music or dance performance. Provide related costumes and props for children to incorporate into their pretend play.
Sing in a group or on their own, increasingly matching the pitch and following the melody.	Play pitch-matching games, humming or singing short phrases for children to copy. Use songs with and without words – children may pitch match more easily with sounds like 'ba'. Sing call-and-response songs, so that children can echo phrases of songs you sing. Introduce new songs gradually and repeat them regularly. Sing slowly, so that children can listen to the words and the melody of the song.
Develop storylines in their pretend play.	Provide a wide range of props for play which encourage imagination. Suggestions: different lengths

Children in reception will be learning to:	**Examples of how to support this:**
	and styles of fabric can become capes, the roof of a small den, a picnic rug or an invisibility cloak. Support children in deciding which role they might want to play and learning how to negotiate, be patient and solve conflicts. Help children who find it difficult to join in pretend play. Stay next to them and comment on the play. Model joining in. Discuss how they might get involved.
Explore and engage in music making and dance, performing solo or in groups.	Notice and encourage children to keep a steady beat, this may be whilst singing and tapping their knees, dancing to music, or making their own music with instruments and sound makers. Play movement and listening games that use different sounds for different movements. Suggestions: march to the sound of the drum or creep to the sound of the maraca. Model how to tap rhythms to accompany words, such as tapping the syllables of names, objects, animals and the lyrics of a song. Play music with a pulse for children to move in time with and encourage them to respond to changes: they could jump when the music suddenly becomes louder, for example. Encourage children to create their own music. Encourage children to replicate choreographed dances, such as pop songs and traditional dances from around the world. Encourage children to choreograph their own dance moves, using some of the steps and techniques they have learnt.

Many thanks to

Dr Julian Grenier, lead of the East London Research School and headteacher of Sheringham Nursery School and Children's Centre, for his work in developing this guidance, alongside the early years organisations, practitioners and professionals who kindly contributed their advice and expertise.

ICAN for giving permission for some of their materials on Communication and Language to be used and replicated within this guidance.

Nicola Burke, author of Musical Development Matters (MDM) and Early Education for giving permission for material from MDM to be used and replicated within this guidance. MDM can be downloaded in full from Early Education: https://www.early-education.org.uk/musical-development-matters

WS - #0077 - 240621 - C0 - 297/210/4 - PB - 9781838150211

The Art of
José Del Nido

Credits

Art Fantastix Select erscheint etwa vierteljährlich bei mg/publishing/, Lochfeldstr. 28c, D-76437 Rastatt.
Internet: www.art-fantastix.de - E-Mail: redaktion@art-fantastix.de

Kunst / *Art*	**José del Nido**
Satz / *Set & Layout*	**Uwe Wallbaum**
Übersetzung (Spanisch/Deutsch) / *Translation (Spanish/German)*	**Sybille Schellheimer**
Übersetzung (Deutsch/Englisch) / *Translation (German/English)*	**Ralf Heinrich & Tim Jiardini**
Chefredaktion / *Editor-in-chief*	**Ralf Heinrich**

mg/publishing/ dankt José del Nido für sein Vertrauen in uns und die angenehme Zusammenarbeit.
mg/publishing/ wpuld like to thank José del Nido for his belief in us and the great cooperation.

Druck / *Printing:* Konradin Druck, Leinfelden-Echterdingen (Germany)
Vertrieb Fachhandel / *Distribution special market:* Modern Graphics Distribution, Lochfeldstr. 3o, D-76437 Rastatt (Germany).
Vertrieb Presse / *Distribution press:* BPV Medien Vertrieb GmbH & Co. KG, Römerstr. 9o, 79618 Rheinfelden(Germany).

PRINTED IN GERMANY

This edition courtesy of Fanfare, P.O. Box 114, Wisbech, PE13 4WF, England.
Co-published in North America by SQP Inc.,38 Rappleyea Rd - Howell, NJ o7731, U.S.A.
Distributed in Italy by Dream Colours srl - Lucca - Italy.

VORWORT

Der vorliegende Band wird den geneigten Leser nicht unberührt lassen. Er präsentiert die Werke des spanischen Künstlers José Del Nido, ein Illustrator, der sowohl für die Werbung als auch für Verlage tätig ist, und der sich seit einigen Jahren mit einer Kunstform auseinandersetzt, die er heute perfekt beherrscht: die digitale Technik im Dienste der Illustration. Auf diesem Gebiet ist Del Nido ein wahrer Meister. Durch die Gestaltung mit Painter und vielen anderen Grafikprogrammen erscheinen uns die Texturen, Hintergründe, Figuren und Szenerien seiner Bilder mit einem unglaublichen Realismus und bilden ein kompaktes und ganzheitliches Werk, das in der endgültigen Illustration seinen Höhepunkt findet.

Was die Thematik betrifft, so ist dieser Band zweifellos eine Hommage an die weibliche Gestalt, an die Sinnlichkeit der Frau in fantastischen und unwirtlichen Welten und in verwüsteten Landschaften, die durch das Böse, die Zauberei und die geheimen Wissenschaften zu wahren Albträume wurden. Bei diesem Abstieg in die schrecklichsten Höllen tritt die Frau – manchmal Opfer, manchmal Henker – als wahre Protagonistin auf. In ihrer ganzen sinnlichen Schönheit zeigt sie sich, bietet sie sich dem Betrachter an.

Diese Auswahl der besten Werke Del Nidos hält uns in ihrem ganz eigenen, fantastischen Universum gefangen. Es ist eine Rückkehr zur besten heroischen Fantasy-Kunst, die in den achtziger Jahren die Zeitschriftencover auf der halben Welt zierte. Damals ging es um eine idealisierte Darstellung der Personen. Del Nido aber modernisiert mit Hilfe der neuesten digitalen Technik das Beste dieses Genres und setzt damit Maßstäbe für Künstler von heute und morgen.

Schließlich wird in dem vorliegenden Band deutlich, dass in jeder Zeichnung Del Nidos viele Stunden Arbeit am Computer stecken, während derer er Farben nuanciert, Schattierungen einfügt und Texturen gestaltet. Auch das kleinste Detail wird nicht vernachlässigt. Der Computer erweist sich als hochentwickeltes Werkzeug, das eine große Hilfe sein kann, aber nichts taugt, wenn die Hand, die es bedient, nicht die Hand eines Künstlers ist.

FOREWORD

The volume in hand will not leave the gentle reader unaffected. It presents the works of the Spanish artist José Del Nido. This illustrator, who has worked in advertising and publishing, has wrestled for years with an art form that today he controls perfectly: the digital technique of illustration. In this field Del Nido is a true master. By using Painter and other graphic design software, the textures, backgrounds, figures and sceneries in his work appear with an unbelievable realism and form a compact and comprehensive piece, which finds its climax in the final illustration.

The theme of this book undoubtedly is homage to the female form, to the sensuality of the woman in fantastic and inhospitable worlds and in wastelands. It is this sensuality that becomes true nightmare through the introduction of evil, dark sorcery and the secret sciences. With this descent to the most horrible hells, the woman – sometimes victim, sometimes executioner – appears as a true protagonist. With all her sensual beauty she reveals herself and offers herself to the viewer.

This selection of the best of Del Nido's work holds the viewer captive in his own particular and fantastic universe. It is a return to the best heroic fantasy art, which in the 80's decorated magazine covers all over the world. At that time the focus was on the idealized representation of the individual. Del Nido modernizes the best of this genre with the help of the latest digital techniques, and by doing so, sets new standards for the artists of today and of tomorrow.

Hours of painstaking work at the computer are put into every illustration, enhancing colors, adding shades and creating textures. Even the smallest detail is attended to. The computer proves to be a very advanced tool, a great benefit when in the hands of an artist.

José Del Nido

BIOGRAPHIE

José Del Nido wurde am 29. Mai 1959 in Cornellà geboren, nur wenige Kilometer von Barcelona (Spanien) entfernt. 1976 begann er eine Ausbildung an der Kunstschule von Cornellà, die damals von Raimon Llort i Gasset geleitet wurde. Hier erlernte er die ersten Mal- und Zeichentechniken. Trotz dieser akademischen Ausbildung eignete er sich den Großteil seiner Techniken autodidaktisch an. Seine ersten Werke zeigte er – sowohl in Gemeinschafts- als auch in Einzelausstellungen – in kleinen Sälen in Barcelona und Umgebung.

Ab 1980 erschienen Del Nidos erste Cover in verschiedenen lokalen Fanzines, und er gestaltete Illustrationen für die Werbekampagnen ortsansässiger Unternehmen. Bald danach begann der spanische Verlag Norma Editorial seine Arbeiten als Illustrator zu veröffentlichen und sie auch außerhalb Spaniens bekannt zu machen. Seit einigen Jahren und auf Anregung von J. Ramón Domingo, einem herausragenden Werbezeichner, mit dem ihn eine enge Freundschaft verbindet, beschäftigt sich Del Nido mit der digitalen Kunst und macht die neue Technologie für sich nutzbar.

Auf diesem Gebiet arbeitet er mit verschiedenen Programmen, von denen besonders Painter zu erwähnen ist. Painter ermöglicht es, Ölfarben, Aquarellfarben, den Airbrush und digitale Farben sowie herkömmliche Malerfarben zu verwenden.

Zur Zeit arbeitet Del Nido als Illustrator sowohl für die Werbung als auch für verschiedene Verlage. Auf dem Gebiet der Fantasy-Illustration steht er weiterhin in enger Verbindung mit Norma Editorial, über den er Cover in mehreren europäischen Ländern, darunter Deutschland und Italien, veröffentlicht hat. Außerdem hat er Arbeiten für Argentinien, Schweden und Portugal angefertigt.

Für zahlreiche Serien hat Del Nido die Cover gestaltet, unter anderen für John Sinclair, Vampira, Proffesor Zamora, Torn, Grusel-Schocker (Bastei), Lanciostory, Skorpio (Euraeditoriale, Italien), Amor y Aventuras, Pesadillas, Fantasville (Ediciones B, Spanien), Romántica (Suma de Letras, Spanien), Panic (Columna, Spanien). Außerdem hat er für Editorial Planeta, Plaza y Janés, Ultramar ed. und andere gearbeitet. Überdies hat er die Werbekampagnen verschiedener Firmen entworfen, wie zum Beispiel Nestlé, Frigo, Continente, Ferrys, Vileda und Eroski. In Zusammenarbeit mit J. Ramón Domingo hat er an Projekten für Terra Mítica, Port Aventura, Enciclopedia Activa Informática und andere mitgewirkt.

Zur Zeit lebt Del Nido in Sant Joan Despí (Barcelona), wo sich auch sein Atelier befindet.

BIOGRAPHY

José Del Nido was born on May 29, 1959, in Cornellà, just a few miles outside of Barcelona, Spain. In 1976 he started his education at the art school of Cornellà, which was run at that time by Raimon Llort i Gasset. It was there where he was exposed to his very first painting and drawing techniques. In spite of his academic education most of his technique was self-taught. He put his first works on exhibition in small halls in Barcelona and the surrounding region – exhibiting with other artists as well as alone.

Beginning in 1980 Del Nido's first covers were published in various local fanzines, and he started creating illustrations for the advertising campaigns of local companies. Soon thereafter the Spanish publisher Norma Editorial began publishing his works as an illustrator, spreading his fame throughout Spain and beyond. For some years now, with encouragement from the extraordinary advertising illustrator, J. Ramón Domingo, a good friend of his, Del Nido has busied himself with digital art, cultivating the new technology for himself.

In the field of digital art, Del Nido works with various software programs, of which Painter stands out foremost. Painter makes it possible to use oil colors, acrylics, airbrush and digital colors as well as conventional colors.

At present Del Nido has been working as an illustrator in advertising and for various publishers. In the field of fantasy illustration he is still linked to Norma Editorial, who published his work on covers in several other European countries beside Spain, among them Germany and Italy. He has also created works for Argentina, Sweden and Portugal.

Del Nido created the covers for numerous series, among them are John Sinclair, Vampira, Proffessor Zamora, Torn, Grusel-Schocker (Bastei, Germany), Lanciostory, Skorpio (Euraeditoriale, Italy), Amor y Aventuras, Pesadillas, Fantasville (Ediciones B, Spain), Romántica (Suma de Letras, Spain), Panic (Columna, Spain). He also worked for Editorial Planeta, Plaza y Janés, Ultramar and others. Furthermore he designed advertising campaigns for various companies, e.g. for Nestlé, Frigo, Continente, Ferrys, Vileda und Eroski. In cooperation with J. Ramón Domingo he played his part on projects for Terra Mítica, Port Aventura, Enciclopedia Activa Informática and others.

Del Nido currently resides in Sant Joan Despí (Barcelona, Spain), where his studio is located.

del Nido

del Nido

del Nido

del Nido

del Nido

del Nido

del Nido

del Nido

del Nido

del

del Nido

del Nido

del Nido

del Nido

del Nido

del Nido

del Nido

del Nido

del Nido

del Nido

del Nido

del Nido

del Nido

del Nido

del Nido

del Nido

del Nido

del Nido

del Nido

del Nido

del Nido

del Nido

del Nido

del Nido

del Nido

del Nido

del Nido

del Nido

del Nido

del Nido

del Nido

del Nido

del Nido

del Nido

del Nido

del Nido

del Nido

del Nido

del Nido

del Nido

del Nido

del Nido

del Nido

del Nido

del Nido

del Nido

del Nido

del Nido

del Nido

del Nido

del Nido

del Nido

INDEX

Bildunterschriften/captions: Clara Edo Comellas

Seite/Page 8

SUBSERVIENCE (1998)

28 x 40 cm/11" x 15,7". Technik: Digital / Painter.
Der Raum wird von der Dunkelheit eingehüllt. Die ungewöhnliche Ruhe verstört die Seele des Mädchens nicht.
The room is enveloped in darkness. The unusual silence does not unsettle the soul of the girl.

Seite/Page 9

THE EMBRACE OF THE DEAD (2000)

30 x 43 cm/11,8" x 16,9". Technik: Digital / Painter.
In den Armen der Toten ruht die Dame mit dem teuflischen Anhänger.
In the arms of the dead rests the lady with the evil pendant.

Seite/Page 10

THE BATH OF THE HOLLOW-HEARTED (1999)

28 x 40 cm/11" x 15,7". Technik: Digital / Painter.
Die Treulose badet in den Wassern der Hölle unter den Augen zweier Wächter.
The faithless takes a bath in the waters of hell, under the eyes of two guardians.

Seite/Page 11

WINGED DEVIL (2000)

30 x 41 cm/11,8" x 16,1". Technik: Digital / Painter.
Das geflügelte Monster hat es geschafft, seine Gefangene auf dem Gipfel des Feuerbergs anzuketten.
The winged monster chains its prisoner on top of the Fire Mountain.

Seite/Page 12

GLAMOUR OF EVIL (1998)

28 x 41 cm/11" x 16,1". Technik: Digital / Painter.
In den Tunneln der Totenstadt entdeckt Andromeda den Totenschädel des Großherrn des Bösen.
In the tunnels of the Dead City, Andromeda discovers the skull of the Master of Evil.

Seite/Page 13

THE LEGACY (1999)

28 x 40 cm/11" x 15,7". Technik: Digital / Painter.
Obwohl die Vampirin sich bereits einen großen Teil des Schatzes angeeignet hat, ist ihr Blutdurst noch nicht gestillt.
Although the vampire has acquired a big part of the treasure, her thirst for blood is unquenched.

Seite/Page 14

THARKAYA (1998)

28 x 40 cm/11" x 15,7". Technik: Digital / Painter.
Eine der Wächterinnen der Nacht steht kurz davor, zu handeln.
One of the Sentinels of the Night is about to act.

Seite/Page 15

SATAN'S DAUGHTER (1999)

28 x 40 cm/11" x 15,7". Technik: Digital / Painter.
Satans Tochter führt ihren Drachen in der Grotte der verlorenen Seelen spazieren.
Satan's daughter takes a stroll with her dragon in the Caverns of the Lost Souls.

Seite/Page 16

THE PATH (1998)

28 x 40 cm/11" x 15,7". Technik: Digital / Painter.
Die Kriegerin des Halbmondes hat das Schloss auf dem zerklüfteten Berg erreicht.
The Warrior of the Half Moon reaches the castle on top of the rugged mountain.

Seite/Page 17

THREE HORNS (1998)

30 x 41 cm/11,8" x 16,1". Technik: Digital / Painter.
Die dreigehörnte Waffe glänzt im Licht des Vollmondes in den Händen ihrer Herrin.
The three horned weapon in the hands of its mistress shines in the light of the full moon.

Seite/Page 18

THE SENTINEL (2001)

30 x 40 cm/11,8"x 15,7". Technik: Digital /Painter.
Die Wächterin des Brunnens ahnt, dass sich jemand nähert.
The Sentinel of the Well is aware that someone is approaching.

Seite/Page 19

FULL MOON (1999)

28 x 40 cm/11" x 15,7". Technik: Digital / Painter.
Mit dem Vollmond erscheint die Vampirin. Trotz ihrer Schönheit – nähere dich ihr nicht!
The vampire appears with the full moon. In spite of her beauty – do not approach her!

Seite/Page 20

ANXIETY (1999)

28 x 40 m/11" x 15,7". Technik: Digital / Painter.
Der Atem des Drachen dringt in ihre Lungen ein, die vor Angst zerreißen. Das Ergebnis wird tödlich sein.
The breath of the dragon invades her lungs, which burst in fear. The result is lethal.

Seite/Page 21

RESURRECTION (1998)

29 x 43 cm/11,4"x 16,9". Technik: Digital / Painter
Ihr fehlt der Atem, ihre Lungen stehen kurz vor dem Zerbersten. Dann entsteigt Ardora als einzige dem Blutmeer.
She runs out of breath, her lungs about to burst. Suddenly Ardora rises from the sea of blood.

Seite/Page 22

SHE IS ALONE (2002)

31 x 42 cm/12,2"x 16,5". Technik: Digital /Painter.
Noch gibt es eine Überlebende.
There is still one survivor.

Seite/Page 23

THE REST (1998)

30 x 44 cm/11,8"x 17,3". Technik: Digital /Painter.
Tief im Wald von Grondar wartet eine Kriegerin darauf, zu handeln.
Deep in the woods of Grondar a warrior is waiting to act.

Seite/Page 24

THE OATH (1998)

29 x 41 cm/11,4"x 16,1". Technik: Digital / Painter.
Unter den Augen ihres Herrn schwört sie Gehorsam. Sie wird an seiner Seite sein, auch wenn sie im Kampf ihre Seele verliert.
Facing her master she swears obedience. She will stay by his side, even if she loses her soul during the battle.

Seite/Page 25

RATS (1999)

28 x 40 cm/11" x 15,7". Technik: Digital / Painter.
Die Herrin der Ratten wird ins Freie treten und...
The Queen of the Rats will step outside and...

Seite/Page 26

RELEASE (2003)

30 x 43 cm/11,8"x 16,9". Technik: Digital / Painter.
Adrehana befreit die drei Geister der Brunnen des Bösen.
Adrehana releases the three ghosts at the Well of Evil.

Seite/Page 27

IN THE DARKNESS (1999)

28 x 40 cm/11" x 15,7". Technik: Digital / Painter.
Venus erwacht aus ihrem Traum in der Finsternis der Welt der Verteufelten.
Venus awakes from her dream in the darkness of the world of the damned.

Seite/Page 28

AURA (2000)

28 x 40 cm/11" x 15,7". Technik: Digital / Painter.
Nach der Meditation entflammt ihre Aura vom Feuer der Rache.
After meditating her aura is aroused by the fires of revenge.

Seite/Page 29

CAPTIVATED (1999)

28 x 40 cm/11" x 15,7". Technik: Digital / Painter.
Die Schergen des Teufels dringen in die Träume der Gefangenen ein.
Devil's henchmen intrude on the dreams of the prisoners.

Seite/Page 30

THE VIRGIN (1999)

28 x 40 cm/11" x 15,7". Technik: Digital / Painter.
Unter den Augen des Unaussprechlichen bereitet sich die Jungfrau auf ihre Opferung vor.
Facing the unspeakable the virgin prepares herself for the sacrifice.

Seite/Page 31

FIGHT TO DEATH (2000)

30 x 40 cm/11,8"x 15,7". Technik: Digital /Painter.
Die Menge bricht in Jubel aus, als die schöne Gladiatorin ihrem ersten Gegner den Tod bringt.
The crowd cheers as the beautiful gladiator brings death to her first opponent.

Seite/Page 32

HOSTAGE (2000)

28 x 40 cm/11" x 15,7". Technik: Digital / Painter.
Dem Tod bleibt noch eine Geisel.
One hostage remains, awaiting death.

Seite/Page 33

THE FANTASTIC FOREST (2001)

30 x 42 cm/11,8"x 16,5". Technik: Digital /Painter.
Jemand hat alles gesehen: die Drachenfrau.
Someone saw everything: the dragon lady.

Seite/Page 34

HOLE TO HELL (1996)

28 x 40 cm/11" x 15,7". Technik: Digital / Painter.
Die Lanze der Macht ist bei den Höllentoren angekommen.
The Spear of Power has arrived at the Gates of Hell.

Seite/Page 35

THE LIGHT (2oo2)

3o x 42 cm/11,8"x 16,5". Technik: Digital /Painter.
Das Licht der Rache erscheint der flüchtigen Prinzessin.
The Light of Revenge appears to the escaping princess.

Seite/Page 36

CHALLENGE (1999)

28 x 4o cm/11" x 15,7". Technik: Digital / Painter.
Die letzte Schlacht wird gegen den mächtigen Herrn der Finsternis geschlagen.
The final battle will be fought with the mighty prince of darkness.

Seite/Page 37

DARK LAKE (1999)

28 x 41 cm/11"x 16,1". Technik: Digital / Painter.
Im Wald lässt die Koboldprinzessin Blut aus dem dunklen See hervorquellen.
In the forest the goblin princess brings forth blood from the Dark Lake.

Seite/Page 38

SEWERS (1996)

28 x 41 cm/11"x 16,1". Technik: Digital / Painter.
Der Zauberin gelingt es, das Glastor zu zerbrechen. Auf ihrer riesigen Ratte reitet sie weiter durch die Kanalisation.
The enchantress succeeds in breaking the Glass Gate. She continues riding through the sewers on the back of her giant rat.

Seite/Page 39

FALSE SKIN (1997)

32 x 4o cm/12,6"x 15,7". Technik: Digital /Painter.
Die wahre Realität kommt auf der Haut zum Vorschein.
The true reality appears on the skin.

Seite/Page 4o

WITNESS (1999)

28 x 4o cm/11" x 15,7". Technik: Digital / Painter.
Nur die Gefangene ist Zeugin des Geschehens.
Only the prisoner is witness to the event.

Seite/Page 41

ZERELA'S WHIP (2oo2)

29 x 42 cm/11,4"x 16,5". Technik: Digital /Painter.
Auf ihren Feuertentakeln trifft die Amazone mit der Zauberpeitsche ein.
The Amazon with the enchanted whip arrives on her fire tentacles.

Seite/Page 42

SEDUCTION (1998)

28 x 4o cm/11" x 15,7". Technik: Digital / Painter.
Sie verführt den Wächter der Finsternis.
She seduces the Guardian of Darkness.

Seite/Page 43

THE ICE TOWER (2oo3)

3o x 43 cm/11,8"x 16,9". Technik: Digital / Painter.
Die besten Waffen für die Eroberung des Eisturms sind schon bereit.
The best weapons are ready for the conquest of the Ice Tower.

Seite/Page 44

THE SPHINX (1998)

29 x 4o cm/11,4"x 15,7". Technik: Digital /Painter.
Die Kriegerin auf dem Adler entkommt dem Zorn der Sphinx mit Hilfe ihres magischen Schwertes.
With help from her magic sword, the warrior on the eagle escapes the wrath of the Sphinx.

Seite/Page 45

POSSESSED (1999)

28 x 42 cm/11"x 16,5". Technik: Digital / Painter.
Mit einem Triumphschrei der Siegerin endet der Albtraum.
The nightmare ends with the victor's shout of triumph.

Seite/Page 46

SIGNS (1999)

28 x 42 cm/11"x 16,5". Technik: Digital / Painter.
Die Blutspuren werden den sich nähernden Heeren den Weg weisen.
The trails of blood reveal the path to the approaching armies.

Seite/Page 47

THE SLAVE (2ooo)

32 x 4o cm/12,6"x 15,7". Technik: Digital /Painter.
Der Geist des Teufels erscheint, um seiner Sklavin eine Mission aufzutragen.
The ghost of the devil appear, sending his slave on a mission.

Seite/Page 48

HARPYIE (2oo1)

32 x 4o cm/12,6"x 15,7". Technik: Digital /Painter.
In den Bergen des Vergessens lebt die Dämonenfrau.
In the Mountains of Oblivion lives the Demon Lady.

Seite/Page 49

SKODEER (2oo1)

28 x 4o cm/11" x 15,7". Technik: Digital / Painter.
Der Schwarze Magier verwendet den Ring der Kobras um die boshafte Skodeer anzurufen.
The Black Magician uses the Ring of Cobras to summon the malicious Skodeer.

Seite/Page 5o

DRAGON BLOOD (2oo2)

3o x 41 cm/11,8" x 16,1". Technik: Digital / Painter.
Eine der Kämpferinnen ruht sich aus, während das Blut des Grünen Drachen noch warm ist.
One of the warriors rests while the blood of the Green Dragon is still warm.

Seite/Page 51

ENTRANCE OF THE TEMPLE (1999)

28 x 44 cm/11"x 17,3". Technik: Digital /Painter.
Nur die Fledermäuse nehmen die Gegenwart einer Fremden am Eingang des Tempels Shaya wahr.
Only the bats perceive the presence of a stranger at the entrance to the Temple of Shaya.

Seite/Page 52

LADY OF THE FUTURE (2oo3)

3o x 41 cm/11,8" x 16,1". Technik: Digital / Painter.
Die Schiffe erreichen die Stadt der Flüchtlinge, während die Schwarze Dame alles aus der Höhe betrachtet.
The ships arrive at the City of Refugees while the Black Lady looks on from the heights.

Seite/Page 53

ENCHANTRESS (1998)

29 x 49 cm. Technik: Digital / Painter.
Die mächtige Zauberin schickt ihre schnellsten Diener, damit sie ihr den Flüchtigen bringen.
The mighty enchantress sends her fastest servant to bring the fugitive to her.

Seite/Page 54

THEY'RE COMING (2oo2)

3o x 41 cm/11,8" x 16,1". Technik: Digital / Painter.
Auf den Mauern des Steinschlosses wartet sie auf die Ankunft der Armeen der Nacht.
On the walls of the stone castle she awaits the arrival of the Armies of the Night.

Seite/Page 55

ALLIES (2ooo)

28 x 4o cm/11" x 15,7". Technik: Digital / Painter.
Mit diesem Bund wird die Welt der Nacht die Kraft sein, welche die Armeen der leeren Seelen beherrscht.
With this alliance the world of the night will have the power to rule the Armies of the Empty Souls.

Seite/Page 56

COBRA (2oo2)

3o x 41 cm/11,8" x 16,1". Technik: Digital / Painter.
Die Rote Kobra und die Schlangenfrau sind bereit zum blutigen Kampf.
The Red Cobra and the Snake Lady are ready for the bloody battle.

Seite/Page 57

CEMETERY (2ooo)

28 x 4o cm/11" x 15,7". Technik: Digital / Painter.
Der Vollmond erleuchtet den Friedhof, auf dem die Zauberin ihre Opfer sucht, um ihren Rachedurst zu stillen.
The full moon illuminates the cemetery where the sorceress is looking for victims to quench her thirst for revenge.

Seite/Page 58

AURORA (1998)

28 x 42 cm/11" x 16,5". Technik. Digital / Painter.
Das Schwert der Macht liegt bereits in den Händen der Königin der Morgenröte.
The Sword of Power is in the hands of the Queen of the Dawn.

Seite/Page 59

DESPERATION (1997)

31 x 4o cm/12,2" x 15,7". Technik: Digital/Painter.
In den Sümpfen der Verzweifelung leben Schlangen und eine Wächterin der Finsternis.
In the Swamps of Desperation live snakes and a Sentinel of the Darkness.

Seite/Page 6o

THE BLACK MOUNTAIN (1997)

28 x 4o cm/11" x 15,7". Technik: Digital / Painter.
In den Höhlen des Schwarzen Berges wartet Neros Tochter auf den Drachen der Tiefen.
In the caves of the Black Mountain Nero's daughter waits for the Dragon of the Abyss.

Seite/Page 61

PATROL (2oo2)

3o x 41 cm/11,8" x 16,1". Technik: Digital / Painter.
Die geflügelten Teufel fliegen über die Hügel, während eine Amazone ihren Rundgang macht.
The winged devils fly over the hills while an Amazon is on patrol.

Seite/Page 62

LADY OF THE LAKE (2ooo)

28 x 4o cm/11" x 15,7". Technik: Digital / Painter.
Die fünf Totenschädel verlassen den Körper der Herrin des Sees.
The five skulls leave the body of the Queen of the Lake.

Seite/Page 63

QUEEN SKULL (1998)

28 x 4o cm/11" x 15,7". Technik: Digital / Painter.
Die Königin Skull mit ihren drei Kriegern erscheint inmitten des tödlichen Nebels der Höhlen ihres Königreiches.
Queen Skull appears with her three warriors in the middle of a lethal fog in the caverns of her realm.

Seite/Page 64

CRIMSON CAVES (1998)

28 x 4o cm/11" x 15,7". Technik: Digital / Painter.
In den purpurnen Grotten versteckt sich die weiße Dame. Sie wird nur begleitet von sechs Totenschädeln, die sie vor dem Guten bewahren.
In the Crimson Caves the White Lady hides. She is accompanied by six skulls, protecting her from the forces of good.

Seite/Page 65

MIGHTY SWORDS (1998)

28 x 4o cm/11" x 15,7". Technik: Digital / Painter.
Endlich konnte sich die Pferdefrau die Schwerter der Macht aneignen.
Finally the Horse Woman acquires the Swords of Power.

Seite/Page 66

DESTRUCTION (2oo2)

3o x 41 cm/11,8" x 16,1". Technik: Digital / Painter.
Teuflische Maschinen, Asphaltreiter, versuchen, aus den Städten zu fliehen, die durch die Gier der Mächtigsten zerstört wurden.
Devilish machines and Asphalt Riders try to escape from the cities, which were destroyed by the greed of the rulers.

Seite/Page 67

CONFRONTATION (1999)

28 x 4o cm/11" x 15,7". Technik: Digital / Painter.
Ein Wutausbruch der Grottenfrau bringt dem letzten Kerkermeister den Tod.
An angry outburst from the woman of the caverns brings death to the last jailer.

Seite/Page 68

UNEXPECTED (1998)

28 x 4o cm/11" x 15,7". Technik: Digital / Painter.
Der Troll der grünen Seele, der das Nebeltal bewohnt, befindet sich schon hinter ihr. Ihre Reaktion kommt wohl zu spät...
The Troll of the Green Soul, who lives in the foggy valley, is already behind her. Apparently her reaction is too late...

Seite/Page 69

THE WATER OF THE LAKE (2oo1)

28 x 4o cm/11" x 15,7". Technik: Digital / Painter.
Der Drache der schwarzen Dame stillt seinen Durst an den Wassern des Sees, in dem noch immer die Seelen der Verdammten umherirren.
The dragon of the Black Lady quenches its thirst at the waters of the lake, where the souls of the damned still wander.

Seite/Page 7o

WISHING WELL (2oo1)

28 x 4o cm/11" x 15,7". Technik: Digital / Painter.
Die Vampirin tritt aus den Nebeln des Wunschbrunnens.
The vampire steps out of the mist at the Wishing Well.

Seite/Page 71

MIGTHY GHOST (1999)

29 x 4o cm/11,4"x 15,7". Technik: Digital /Painter.
Eine maskierte Amazone benetzt die große Säule der Mutlosigkeit mit Blut und schwört damit den Zorn des mächtigen Geistes herauf.
A masked Amazon sprinkles the Giant Pillar of Despondency with blood, thus bringing the wrath of a mighty ghost.

Seite/Page 72

NIGHTMARES (2oo1)

28 x 4o cm/11" x 15,7". Technik: Digital / Painter.
Schreckenerregende Bilder nähern sich der Zauberin, die von unruhigen Träumen geplagt wird. Vielleicht sind sie Teil ihrer Albträume - vielleicht auch nicht...
Horrible visions plague the sorceress, tormenting her restless sleep. Maybe they are part of her nightmares - maybe not...

Seite/Page 73

THE EMISSARY (2oo2)

29 x 42 cm/11,4"x 16,5". Technik: Digital /Painter.
Nach dem erbitterten Kampf, indem sie sich ihre Wunden zugefügt hat, wartet Aranha, Prinzessin der Finsternis, auf den Boten Satans, während der Wald noch in Flammen steht.
After the fierce battle, Aranha, Princess of the Darkness, waits for Satan's emissary, with the forest still in flames.

Seite/Page 74

OFFERING (2oo2)

29 x 42 cm/11,4"x 16,5". Technik: Digital /Painter.
Die Mönche der Gemeinschaft des Bösen Wortes binden die Jungfrau an die Säule des grünen Auges um das Opfer zu vollziehen.
The monks of the Brotherhood of the Evil Word tie the virgin to the Pillar of the Green Eye, in preparation for the offering.

Seite/Page 75

THE RETURN (1997)

3o x 42 cm/11,8"x 16,5". Technik: Digital /Painter.
In der Geisterwüste sucht eine Kriegerin den Weg zurück ins Verlorene Tal.
In the Vulture's Desert a warrior searches for the way back to the Lost Valley.

Seite/Page 76

SPLIT TONGUES (2ooo)

28 x 4o cm/11" x 15,7". Technik: Digital / Painter.
Erregt durch den Geruch von Angst, den ihre Beute ausströmt, nähern sie sich, bereit, den Opferritus zu vollziehen.
Aroused by the smell of fear given off by the offering, they approach, ready to perform the rite of sacrifice.

Seite/Page 77

TATOO (2oo1)

29 x 41 cm/11,4"x 16,1". Technik: Digital / Painter.
Tatoo wurde von der grünen Riesenspinne gefangengenommen.
The giant green spider captures Tatoo.

Seite/Page 78

UNINHABITED KINGDOM (2oo2)

3o x 41 cm/11,8" x 16,1". Technik: Digital / Painter.
Auf dem Thron des Schlosses des Unbewohnten Königreiches liegt tot ihr Herr, während Prinzessin Skull die Ankunft des Drachen Smorfo erwartet.
Her master lies dead on the throne of the Castle of the Uninhabited Kingdom, while Princess Skull awaits the arrival of dragon Smorfo.

Seite/Page 79

ZOMBIE (1998)

28 x 42 cm/11"x 16,5". Technik: Digital / Painter.
Tote, die bei Vollmond verführen.
The dead seduce by the light of the full moon.

Seite/Page 8o

DESECRATION (1997)

29 x 4o cm/11,4"x 15,7". Technik: Digital /Painter.
Sie hat das geheime Grab des ersten Vampirs in der Gruft auf dem Friedhof der Wasserspeier entdeckt.
She discovers the secret grave of the first vampire in a tomb at the Cemetery of the Gargoyles.

Seite/Page 81

MOUNTAIN OF THE DEATH (1999)

28 x 4o cm/11" x 15,7". Technik: Digital / Painter.
Der Berg der Toten, der von der Wächterin der Schatten bewacht wird. Auch sie ist tot.
The Mountain of the Dead is guarded by the Sentinel of the Shadows. She is dead, too.

Seite/Page 82

EXPERIMENT (2oo2)

29 x 41 cm/11,4"x 16,1". Technik: Digital / Painter.
Ein verhängnisvolles Experiment. Die Macht der Bestie läuft durch Nabelschnüre, die mit der Maschine des Res verbunden sind.
A fatal experiment. The power of the beast runs through umbilical cords linked to the Machine of Res.

Seite/Page 83

THE CAVE OF THE BROMS (2oo2)

29 x 41 cm/11,4"x 16,1". Technik: Digital / Painter.
Broms' Sklavin, die sich seinen lüsternen Wünschen fügen muss, wartet auf die Gelegenheit, den Schlag zu versetzen, der ihre Rache vollenden wird.
Broms' slave, who must submit to his lewd wishes, waits for a chance to strike out and complete her revenge.

Seite/Page 84

FINAL BREATH (2oo2)

28 x 4o cm/11" x 15,7". Technik: Digital / Painter.
Der Drache stößt seinen letzten Schrei aus. Die weiße Zauberin fängt das Blut auf, das aus seinem Maul fließt.
The dragon expels his final scream. The White Sorceress collects the blood pouring out of its mouth.

Seite/Page 85

LETHAL WOUND (2oo1)

29 x 41 cm/11,4"x 16,1". Technik: Digital / Painter.
Unter den Augen der bedächtigen Marhenia, Göttin der Berge, ringt General Dark mit dem Tode, getroffen von seinem eigenen Schwert.
Facing the thoughtful Marhenia, Goddess of the Mountains, General Dark wrestles with death, and is struck by his own sword.

Seite/Page 86
CAUTION! (2000)
28 x 40 cm/11" x 15,7". Technik: Digital / Painter.
Lass dich nicht von ihrer Schönheit verzaubern! Du könntest der nächste sein...
Don't be fooled by her beauty! You could be next...

Seite/Page 87
DEPTH (2003)
30 x 42 cm/11,8" x 16,5". Technik: Digital/Painter.
Tief im Innern der Festung lebt die Kriegerin mit dem traurigen Blick, die einzige Herrin der Lanze des Greifs.
Deep inside the fortress lives the warrior with the sad face, the only master of the Spear of the Griffin.

Seite/Page 88
ENTRANCE TO HELL (2001)
28 x 40 cm/11" x 15,7". Technik: Digital / Painter.
Das Eintreten ist nicht schwierig. Das Herauskommen ist unmöglich.
Getting in is easy. Getting out is impossible.

Seite/Page 89
BAD COMPANY (2003)
29 x 41 cm/11,4"x 16,1". Technik: Digital / Painter.
Er verspricht ihr, bei der Eroberung des Königreiches der Schwarzen Schatten behilflich zu sein, aber seine eigentliche Absicht ist es, sich ihres Willens zu bemächtigen.
He promises to help her conquer the Kingdom of the Black Shadows, but his real intention is to take hold of her will.

Seite/Page 90
HALLEN'S SWAMP (2003)
29 x 41 cm/11,4"x 16,1". Technik: Digital / Painter.
Seine Wasser lassen Seelen verschwinden. Nur die Verzweifelten baden sich in ihm.
Its waters make souls disappear. Only the desperate take a bath in it.

Seite/Page 91
BLOODBATH (2001)
28 x 40 cm/11" x 15,7". Technik: Digital / Painter.
In Blut gebadet taucht der Körper der Jungfrau auf.
Soaked in blood the body of the virgin appears.

Seite/Page 92
TATTOOED DRAGON (2001)
28 x 40 cm/11" x 15,7". Technik: Digital / Painter.
Die Tätowierung auf der Schulter ist das Zeichen, auf das der Drache gewartet hat. Jetzt weiß er, wer die Frau ist, die vor ihm steht...
The tattoo on her shoulder is the sign the dragon was waiting for. Now he knows who the woman in front of him is...

Seite/Page 93
PROJECTIONS OF THE SKULL (1998)
28 x 40 cm/11" x 15,7". Technik: Digital / Painter.
Riesige Fledermäuse bewachen die Gefangenen auf dem Vorsprüngen des Totenschädels.
Giant bats guard the prisoners on the ledges of the skull.

Seite/Page 94
THE FORGOTTEN SWAMP (2003)
30 x 41 cm/11,8" x 16,1". Technik: Digital / Painter.
Die Bewohner des Sumpfes des Vergessens wissen, dass ihre Herrin darauf wartet, verführt zu werden, von dem, den sie fürchtet und zugleich verehrt.
The inhabitants of the Forgotten Swamp know that their lady waits to be seduced by the one who fears her, but also worships her.

Seite/Page 95
THE POTION (2001)
28 x 40 cm/11" x 15,7". Technik: Digital / Painter.
Vom Hexer in die Enge getrieben wird sie versuchen, das Gebräu, das er ihr anbietet, nicht zu trinken, denn sie weiß, dass es die Essenz der Seelen enthält, die keine Ruhe finden.
With her back to the wall she will try not to drink the sorcerer's brew, because she knows it contains the essence of the souls who never rest.

Seite/Page 96
GRAY WOLF (2001)
28 x 40 cm/11" x 15,7". Technik: Digital / Painter.
Mit ihrem smaragdgrünen Schwert bewaffnet und in Begleitung des Grauen Wolfes, wird sie ihr Revier vor den Fremden verteidigen, die versuchen, ihre Herrschaft zu stören.
Armed with her emerald green sword and accompanied by the Gray Wolf she will defend her territory against the strangers who try to spoil her reign.

Seite/Page 97
THE SHELTER (2000)
28 x 40 cm/11" x 15,7". Technik: Digital / Painter.
Ihre Zuflucht ist der Friedhof. Hierher kommt sie nach ihren nächtlichen Jagden.
Her shelter is the graveyard. She comes here after nocturnal hunting.

Seite/Page 98
LUCIFER'S WIVES (2002)
29 x 41 cm/11,4"x 16,1". Technik: Digital / Painter.
Der Tag wird kommen, an dem sie aus der Finsternis treten und sich unter uns mischen werden. Vielleicht sind sie bereits hier.
The day will come when they will step out of the darkness and mingle with us. Perhaps they are already here.

Seite/Page 99
ENERGY (2003)
30 x 41 cm/11,8" x 16,1". Technik: Digital / Painter.
Die Energie in seinen Eingeweiden steht kurz davor, aufzutauchen und dabei alles einzuebnen, was sich in ihrem Weg befindet. Auch wenn das Alien stirbt, wird sich seine Gattung weiter fortpflanzen.
The energy inside is about to burst forth, destroying everything in its path. The Alien might die, but its species will continue reproducing.

Seite/Page 100
NEON (2003)
30 x 41 cm/11,8" x 16,1".Technik: Digital / Painter.
Alles ist blau, ihre Kraft ist blau, ihre Augen sind blau. Das Blau, das von der Kraft des Neons ausgeht, wenn sie in Berührung mit den Wassern des Sees Eras kommt.
Everything is blue; her power is blue; her eyes are blue. The blue originates from the Power of the Neon, only when she comes in contact with the waters of Lake Eras.

Seite/Page 101
THE EMPRESS (2003)
30 x 41 cm/11,8" x 16,1". Technik: Digital / Painter.
Im Zauberbuch steht alles geschrieben. Der Stab der Zauberin muss nur das Rätsel lösen, welches ihr die Macht geben wird, das Reich des sechseckigen Steins zu regieren.
It is all in the grimoire. The Rod of the Sorceress must solve the riddle, and thus give her the power to rule the Realm of the Hexagonal Stone.

Seite/Page 102
SPIDER GODDESS (2003)
30 x 41 cm/11,8" x 16,1". Technik: Digital / Painter.
Mit ihrer Willenskraft wird die Spinnengöttin den Geist der großen Tarantel anrufen.
Through her willpower the Spider Goddess will summon the ghost of the giant tarantula.

Seite/Page 103
LONG JOURNEY (2003)
30 x 41 cm/11,8" x 16,1". Technik: Digital / Painter.
Ihre Reise nähert sich dem Ende. Niemand erwartet sie. Sie weiß, dass ihre Ankunft Schmerz bereiten wird.
Her journey comes to an end. Nobody expects her. She knows her arrival will cause pain.

Seite/Page 104
DRAGON LADY (2002)
28 x 40 cm/11" x 15,7". Technik: Digital / Painter.
Warum versuchst du, meine Aufmerksamkeit zu erregen?
Why are you trying to attract attention?

Seite/Page 105
THE MIRROR (2003)
30 x 41 cm/11,8" x 16,1". Technik: Digital / Painter.
Auch wenn der Spiegel des Drachen dir zeigt, dass sich dein Aussehen verändert hat, wird deine Seele weiterhin dunkel bleiben.
Although the Mirror of the Dragon shows your face has changed, your soul will stay dark.

Seite/Page 106
HALF MOON (2003)
30 x 40 cm/11,8"x 15,7". Technik: Digital /Painter.
Die Gefangene des Halbmondes lässt ihr Blut auf den Grund des Brunnens der Verzweiflung rinnen.
The Prisoner of the Half Moon lets her blood run to the ground at the Well of Desperation.

Seite/Page 107
BLACK TEARS (2003)
30 x 41 cm/11,8" x 16,1". Technik: Digital / Painter.
Schwarze Tränen fließen über ihr Gesicht. Nur der Mond ist Zeuge ihres bitteren Schmerzes.
Black tears run down her face. Only the moon is witness to her bitter pain.

José Del Nido